Essays in Political Economy and International Public Finance

FINANZWISSENSCHAFTLICHE SCHRIFTEN

Herausgegeben von den Professoren
Konrad, Krause-Junk, Littmann, Oberhauser, Pohmer, Schmidt†

Band 119

PETER LANG
Frankfurt am Main · Berlin · Bern · Bruxelles · New York · Oxford · Wien

Áron Kiss

Essays in Political Economy and International Public Finance

PETER LANG
Internationaler Verlag der Wissenschaften

Bibliographic Information published by the Deutsche Nationalbibliothek
The Deutsche Nationalbibliothek lists this publication in the Deutsche Nationalbibliografie; detailed bibliographic data is available in the internet at http://dnb.d-nb.de.

Zugl.: Berlin, Freie Univ., Diss., 2009

D 188
ISSN 0170-8252
ISBN 978-3-631-59676-0

www.peterlang.de

To my parents

Contents

Acknowledgements

I would like to thank my supervisor, Kai A. Konrad, for his guidance through these years. I would also like to thank my second advisor, Helmut Bester, for his help and comments on the project.

My colleagues at Social Science Research Center Berlin (WZB) provided a motivating and stimulating atmosphere and commented at various stages on each of these chapters. I'd like to thank Johannes Münster, Beate Jochimsen, Sebastian Kessing, Robert Nuscheler, Ashish Chaturvedi, Benny Geys, Florian Morath, Salmai Qari and Margarita Kalamova. I have learned a lot from all of them. Additionally, I'd like to thank Margarita for putting up with having to drink so many cups of tea while listening to what happenned to my papers and me on a given day. The WZB and Free University Berlin provided great working conditions. I would like to thank the administrative staff in both institutions and, in particular, to Babette Hagemann at WZB for helping me solve many bureaucratic puzzles.

I had the opportunity to discuss some of my ideas with guests on research visit to WZB. I am thankful for inspiring conversations with Amihai Glazer, Arye Hillman, Dan Kovenock, Wolfgang Leininger, Ronnie Schöb and Oz Shy. The German Science Foundation's (DFG) project SFB TR-15 allowed me to visit meetings of a network of distingushed researchers. I am thankful for interesting comments on one of my papers by Dalia Marin, Klaas Staal and Konrad Stahl.

I profited from many interesting conversations in formal or less formal surroundings and received a number of helpful suggestions from colleagues at the multitude of Berlin universities and research institutes. I would like to express my gratitude in particular to Daniel Krähmer, Roland Strausz, Jonathan Beck, Joe Clougherty, Tomaso Duso, Ela Glowicka, Michal Grajek, Susanne Prantl, Jo Seldeslachts, Timm Bönke, Frank Neher, Justine Röhmel, Philipp Engler, Runli Xie and Martin Uebele.

During my studies, many professors shaped my personality and views on economics. I am particularly indebted to Elmar Wolfstetter, Michael Burda, Harald Uhlig, Werner Güth, Helmut Lütkepohl and Bengt-Arne Wickström at Humboldt University Berlin and Peter Gedeon at the Budapest University of Economic Science.

On a personal note, there are a lot of people who made my life warmer, more colorful and louder in the hours I wasn't trying to calculate or compose and helped me recharge after each day and each week. I am infinitely indebted to them and I hope they know it even if I do not name them all. But I can't help thanking Julia, Gediminas, Alex, Daniel, Dejan, Martin, Runli, Sophie, Sushmita and Wilfried for the good-old-days moments; Martina, Francesca and Mareen for making me feel at home in Osloer Str.; Marc, Carles, Paul and Hanna for the loud part; András, Gábor and Dávid for welcoming me back on visits to Budapest.

Finally, I'd like to thank my sister Flóra who made the whole journey possible. And Erzsi for her patience, her impatience and all the lessons we learned.

Chapter 1

Introduction

The present thesis is a collection of three essays in Political Economy and International Public Finance. They are bound together by two unifying principles, one relating to the subject matter and one relating to the form. The common principle behind the subject matter of the essays is the preoccupation with the role and functioning of the state. The common principle of 'form' is the method of investigation: All essays use the tools and insights of game theory. The essays cover two topics in this broader field: Chapters 2 and 3 deal with problems of political accountability (Political Economy) while Chapter 4 studies repeated tax competition between states (International Public Finance).

This introductory chapter begins with a sketch of the broad field the essays fit into: the game theoretic analysis of the state. Then, it provides a more in-depth introduction to the two topics analyzed in the three essays, discussing the existing literature and describing how the thesis contributes to the study of political accountability and tax competition.

1.1 Game-theoretic analysis of the state

All three essays of the thesis are 'game theoretic analyses of the state.' This section sketches the development of such an approach and discusses its relationship with related disciplines, labels and schools of thought.

Game theory, the method of investigation, is a set of tools designed to analyze strategic interaction – situations in which what agents prefer to do depends on what other agents do (as opposed to pure choice problems where agents make decisions facing given circumstances of 'Nature' or 'Chance'). While students of human relationships in general – and politics in specific – have always thought about strategic interaction, some results of modern mathematics were needed before game theory in its present form could be developed. The birth of the new methodology can be dated with precision to the publication of a monography by von Neumann and Morgenstern (1944), a mathematician and an economist.

The early age of the game theoretic analysis of the state (roughly the 1950's and 1960's) brought the publication of several path-breaking monographies that defined the research questions and methods for decades to come. Some of these works (e.g. Schelling 1960; Buchanan and Tullock 1962; Riker 1962) applied and extended game theory in new domains of economics and politics, contributing to the spread of its influence in the social sciences. Downs (1957), while not using new game theoretic methods explicitly (applying rather concepts from economics preceding formal game theory, i.e., Hotelling 1929) analyzed strategic interaction of parties in a democracy in a way present day readers would classify as game theory. Further important contributions to the formal analysis of voting and preference aggregation (e.g. Arrow 1951; Black 1958), while based on the theory of choice rather than the theory of games, were influential for the later development of the game theoretic analyses of voting.

In the three decades following the 'founding fathers' the economic study of the state underwent a 'game theoretic turn' similar to that in other fields of economics (e.g. Industrial Organization). During the same period the economic study of the state was marked by a stark dichotomy. On the one hand classical Public Finance concentrated on the existence of market failures and asked what kind of state intervention could solve them. The paradigm of this approach is the description of the notion of externalities and the recommendation to correct them by an appropriate tax (Pigou, 1912). On the other hand the Public Choice school, following Buchanan and Tullock (1962; and other works by the same authors) concentrated on a question neglected by traditional Public Finance: What kind of policies are likely to emerge from state intervention given that the political sphere (just like the market place) is populated with individuals following their own goals and interests? Public Choice theorists argued that there is no reason to think that the state would always do the 'right thing,' which gave emergence to 'government failure,' a concept parallel to market failure.

Although in economic debates arguments along these lines are still contrasted, the borders of traditional (Pigouvian) Public Finance and the more skeptical Public Choice have blurred to a great extent in the last two decades. Market failure and government failure are both parts of the profession's vocabulary. The fields of Public Finance (or Public Economics or Welfare Economics) and Political Economy (the broader discipline encompassing the Public Choice school) are seen as complements or even parts of the same discipline. The essays of this thesis are part of this complex field. Independent of labels, they contribute to a 'game theoretic analysis of the state.'

1.2 The study of political accountability

The study of political accountability is the study of how the prospect of reelection shapes the behavior of politicians in office. Before turning to the contribution of the thesis to the study of political accountability, this section surveys some aspects of the development of the field.

1.2.1 History and previous results

One way of placing the study political accountability into the context of research on elections in general is to place it in a taxonomy of politician motivation and voter motivation. The theory of political accountability is based on the notion that politicians value reelection. This, to be sure, does not contradict the existence of other motives like the wish to shape policies or leave a legacy. In fact, a politician might want to be reelected for these reasons. But for the mechanism of political accountability to work, it is a neccessary condition that they value reelection; and we may leave open the question about the ultimate reasons. Looking at voter motivation, the theory of political accountability views voting as a rational and retrospective act.

Rationality. The game theoretical analysis of politics generally views voters as active players making choices to further their interests based on the available information. While most of this literature can be brought under the umbrella of 'rational voting', approaches introducing bounded rationality or aspects of irrationality have been proposed as well (e.g. 'bandwagon effects' by Hong and Konrad (1998)). Still, modeling voter choice explicitly should be viewed as a choice about Method rather than Substance. Other approaches to similar effect are also conceivable. For example, one could treat voter behavior as a mechanical response to policy outcomes, calibrated perhaps to reflect the empirical relationship between macroeconomic outcomes and election results.[1] Many of the fundamental mechanisms and dilemmas of political accountability would be left unchanged by this methodological decision. The approach taken here reflects the tendency in modern economics to treat all participants of economic or political interactions as agents trying to act optimally. This tendency is best exemplified by the insistence in modern macroeconomics to explain macroeconomic phenomena with theories explicitly relying on 'micro-foundations.'

---

[1]For surveys of the literure on 'economic voting' see Nannestad and Paldam 1994; and Lewis-Beck and Paldam 2000. For further references, see Section 2.2.

Retrospective voting. An early debate among some political scientists and economists was centered around the question whether voting should be viewed as a prospective or a retrospective act. Downs (1957) based his theoretical description of elections on the premise that rational voting must be prospective. Key (1966), on the other hand, found empirical support for retrospective voting. The economic research of recent years reconciled these views to some extent. In the work of Rogoff and Sibert (1988) and Rogoff (1990), actions of the incumbent government are signals about its competency. When voters make their decision about reelection based on a rational and prospective calculus, they should look for the relevant information in the past actions and thus vote retrospectively. Past actions play a structurally similar role in the theory of 'career concerns' (Persson and Tabellini, 2000), which was first developed in the context of organization theory by Holmström (1982b) and Dewatripont et al. (1999a,b). Surveys of the problems related to retrospective versus prospective voting are provided by Drazen (2000) and Persson and Tabellini (2000).

Accountability among the theories of rational retrospective voting. Rational voting can thus be retrospective. Each of the theories described above referred to situations where voters look at past actions to infer something about the 'type' of the politician. This is, in other words, a selection motive: voters would like to have the right kind of politician in power, where the 'right kind' may refer to ability, congruence of preferences with the electorate, or some other factor. The motive for political accountability is different from the selection motive. Voters want to reward politicians for good outcomes and punish them for bad ones to create incentives for them before the fact. If politicians know that their reelection depends on the outcomes under their tenure, they have an incentive to steer those outcomes according to the electorate's preferences. Paradoxically, while creating incentives for politicians is a rational and prospective motive, on election day (i.e. 'ex-post') voting for accountability is a purely retrospective act: it is not direcly related to tomorrow's well-being like it is in the case of the selection motive.

Theoretical insights. The game theoretic study of political accountability was initiated by Barro (1973) and Ferejohn (1986). In the core of these models lies a conflict of interest between the electorate and a politician. They show that the prospect of reelection can induce the politician to act in the public interest. Barro (1973) analyzed how constitutional provisions like the frequency of elections affect the incentives of the politician. Ferejohn (1986) realized that the analysis of political accountability is the application of 'moral hazard' in politics (that is, a principal-agent problem where the voters are the principals and the politician the agent) and analyzed the political accountabiltiy problem in an infinite-horizon setting. A simple exposition of the political accountability problem reduced to its contract-theoretic core is provided by Persson and Tabellini (2000).

The study of political accountability has been extended by an 'adverse selection' element by Austen-Smith and Banks (1989) and Banks and Sundaram (1993); their models consider the possibility that there are different 'types' of politicians. The most important insight of this branch of literature is that voters not only want to 'discipline' politicians (provision of incentives) but also choose the 'right type' of politician (selection).[2] Besley (2006) analyzed such a setup extensively and underlined an interesting trade-off facing voters. Consider a world with two types of politicians: congruent ones (who share the preferences of the electorate) and non-congruent ones (who have different preferences). The time-horizon is two periods: the incumbent politician faces elections after the first period. After the second period the game ends (this is a simplifying assumption having a similar effect as a 'term limit' allowing politicians to run for the same office only twice). If the incentive function of elections works perfectly, then opportunistic politicians will act in the first period like congruent ones would and therefore get reelected. In this case the selection function of elections is compromised: second-term opportunistic politicians act in their own interest. Conversely, if the selection function of elections works perfectly, the electorate can tell apart different types of politicians and only congruent types get reelected. The price of this is that opportunistic types, in the first period, do not restrain themselves.

Recent research related the theoretical study of political accountability to institutions and constitutional choices of democracies. Persson et al. (1997) analyze the principle of separation of powers in the context of political accountability. They consider a political system where there are two politicians (or elected bodies) elected separately, much like the president and the legislature in a presidential system. Both the presi-

[2]The selection motive in these political accountability models is clearly related to the signaling and 'career concern' models described above. As a difference to those, the politicians here do not differ in competency but rather in their motivation.

dent and the legislature makes a decision that contributes to policy – for example the determination of the budget. The authors investigate what constitutional rules about the decision-making procedure allow the voters to attain good government performance through reelection incentives. The result is that political accountability works if both politicians have to agree on a certain proposal; while it breaks down if both politicians can influence the outcome independently from each-other (in this case the 'common-pool' problem emerges). This result allows the authors to analyze formal procedures (e.g. two-stage budgeting) in the context of political accountability.

In an other recent study dealing with constitutional choices and accountability, Maskin and Tirole (2004) consider three simple institutional rules: direct democracy; accountable representatives ('politicians'); and unaccountable representatives ('judges'), and ask which is best under various circumstances. The representatives are better informed than the public about which of two policies are best for the electorate and may or may not share the electorate's preferences. They care about the policy as well as about reelection. In this simple framework an interesting drawback of political accountability becomes apparent: 'pandering'. It is possible that a congruent politician chooses the policy that he prefers less and knows is worse for the electorate only because it corresponds to the prior beliefs of the electorate. In such a case, a 'judge' would serve the public interest better than a 'politician.'[3]

[3]Pandering has been analyzed in different context by Morris (2001). For an other instance of bad outcomes following from political accountability, see Dewatripont and Seabright (2006).

A last group of studies, while also introducing questions of constitutional design to the study of political accountability, has a special relevance for this thesis by establishing a connection to fiscal competition, the topic of Chapter 4 (and section 3 of this Introduction), through questions of federalism and (de)centralization. Besley and Case (1995a) analyze 'yardstick competition' in policy-making. The mechanism (identified first in a different context by Shleifer (1985)) is that voters can look at neighboring jurisdictions to learn about unobservable conditions affecting all jurisdictions to see whether their local politicians do the right thing. As a consequence, high taxes may have less serious electoral consequences when they are observed in other places as well. Seabright (1996) describes a possible trade-off involved in decisions about (de)centralization. Centralized policy-making allows a more efficient policy-coordination across jurisdictions in the presence of externalities. The price to pay for this efficiency gain is that the decision maker is less accountable to the individual constituency because their reelection probability depends on many constituencies. Finally, Myerson (2006) argues that a federal structure may improve political accountability, as governors of the regional level have 'both motive and opportunity' to build a reputation acting in the public interest. The presence of the regional level in his framework makes the worst equilibria (a 'consistent frustration of democracy') disappear.

Empirical results. Early measurements of the effect of elections on politician's behavior were provided by the literature on opportunistic political business cycles. The underlying hypothesis, suggested by Nordhaus (1975), states that politicians try to stimulate the economy before elections to improve the public opinion of their govenment. Thus GDP growth is expected to be higher before elections and lower after elections. Starting with Tufte (1978) and McCallum (1978) several researchers tested various implications of the opportunistic business cycle hypothesis (for an overview, see Drazen (2000, Ch. 7). An influential paper of Alesina, Cohen and Roubini (1992) on a panel of OECD economies finds little evidence for higher growth rates before elections, but some evidence for higher inflation rates after elections and higher budget deficits before elections. These results (see also the overview by Alesina, Roubini and Cohen (1997) and the discussion in Drazen (2000)) suggest that while politicians are trying to manipulate policy *instruments* like spending (with monetary policy apparently accommodating these shocks even in the presence of central bank independence), the effect on aggregate macroeconomic *outcomes* like output is too small or too imprecisely timed to be measured accurately against the noise of other macroeconomic processes.

As opposed to the 'macro' level, which the political business cycle literature focused on, another strand of literature tried to empirically identify the effect of elections on politician behavior on the 'micro' level. In a seminal paper Besley and Case (1995b) depart from the observation that many U.S. states have term limits for the governors office, implying that there are governors who by law cannot stand for reelection. If the prospect of reelection did not affect the behavior of politicians, governors who can stand for reelection and those who can't should behave similarly. Conversely, if we observe a difference in the behavior of both groups, we have reason to think that governors change their behavior to enhance their chance of reelection, e.g. building a reputation for supporting certain policies that differ from their most preferred ones.

Besley and Case find that term-limited governors impose higher sales and income taxes (measured as tax revenues per capita) than their not term-limited counterparts. The effects are in the single-digit percent magnitude and statisticially significant, even after controlling for some state demographic controls, and state and year effects. Other variables (like total state expenditure per capita and state minimum wage) are also affected by term limits. The authors also find that party affiliation of the governor matters; in particular, the effect on public finance variables is mostly driven by term-limited Democratic governors, while the legal minimum wage tends to be lower in the last term of Republican governors.

In a related study cited above, Besley and Case (1995a) find evidence for 'yardstick competition' in regional policy-making. While governors' reelection probability decreases with increasing taxation in their state, it increases with increasing taxation in neighboring states. Further evidence is found in the tax-setting behavior of governors: taxes in neighboring states affect home-state taxes only for the subgroup of governors who can run for reelection, but have no effect on home-state taxes if the governor is term-limited.[4]

[4]The large empirical literature that has appeared since Besley and Case (1995a) is surveyed by Brueckner (2003) and Revelli (2005). While term limits provide a very convenient identification strategy, recent studies also use more indirect ways to identify the effect of electoral pressure on politicians, e.g. the margin of their previous victory (see, e.g., Bordignon et al. (2003); Solé-Ollé

1.2.2 Contribution of the thesis

Two chapters of this thesis study questions related to political accountability, extending the literature in similar ways: Both chapters put the principal-agent relationship between voters and elected politicians in a more detailed institutional context. The context may include details on the electoral system (proportional or majoritarian) and the party system (two-party or multi-party system). Additionally, voter heterogeneity may be taken into account explicitly in the form of dividing voters into ideologically committed partisan voters and independent ('swing') voters; It is independent voters who make political accountability operative. The institutional detail allows us to extend the reach of the study of political accountability to some questions related to party politics, ideology and polarization.

Coalition politics and accountability. Chapter 2 asks a question relevant in proportional electoral systems, as these tend to support a party system with many parties and a high frequency of coalition governments.[5] Do coalition governments have an accountability deficit relative to single-party governments?

The question is closely related to a hypothesis formulated in studies of empirical public finance and political economy. As formulated in the seminal paper by Roubini and Sachs (1989), the 'weak government hypothesis' says that coalition governments are more prone to accumulat-

(2003); Allers and Elhorst (2005); Geys (2006)). See also the discussion below of the related empirical literature on tax competition.

[5]This 'law' was established in political science by Duverger (1954) and further analyzed by Taagepera and Shugart (1989).

ing public debt than single-party governments.[6] More generally, the hypothesis implies that coalition governments tend to produce inefficient outcomes. Roubini and Sachs proposed several possible reasons why the hypothesis might be true, among them the collective action problem (or 'common pool' problem) in the spending of public funds, the lack of commitment power of coalition partners, and the high number of veto players. All these explanations fall in the category of 'post-election politics': they do not take into account the influence of elections. Putting the problem in the context of political accountability, Chapter 2 investigates the significance of elections for the weak government hypothesis.

The chapter's contribution to the study of political accountability is that it introduces the possibility that government consists of multiple decision makers into a simple framework in the style of Persson and Tabellini (2000, Ch. 4).[7] In this way it analyzes coalition government as a team production, a problem of moral hazard in teams. The team of agents (coalition government) produces a team good (government output) while the principal (the representative voter) tries to give it incentives by the prospect of reelection.

Two main results emerge from the analysis. First, coalition government as such does not hamper electoral accountability. Coalition governments can be given appropriate (collective) incentives as long as, in

[6]For a recent review of the literature since Roubini and Sachs, see Ashworth et al., 2005.

[7]In a paper discussed above, Persson et al. (1997) also analyze the political accountability of two politicians. In that framework, however, the politicians are reelected separately, rather than being part of the same body.

the presence of an electoral alternative, the coalition as a whole can be voted out of power. The reason is that reelection conditional on government performance works like a *discrete team bonus*, the type of contract between the principal (voters) and the team of agents (coalition government) that was shown by Holmström (1982a) to solve the moral hazard problem in teams.

The second main finding is that the accountability of a coalition government becomes problematic when, in the absence of a real electoral alternative, the government cannot be removed as a whole. In this case coalition parties can not be given appropriate team incentives. This case is called the 'unity government.' With the unity government in place, voters do not decide between incumbent and challenger but rather between two political forces, both of which are in power. How to give incentives to government or, in other words, who to vote for after good or bad outcomes is the problem facing the voters.

There are two further assumptions that play an important role in the analysis. First, voters can only observe the sum of efforts by coalition parties rather than each individual effort separately. Second, costly effort reducing government performance ('sabotage') is possible.The approach taken by the thesis is to solve for the best outcome the voters can obtain in the framework of the simple accountability game. In the optimum, voters make one of the coalition parties responsible for the outcome. They vote for that party if government performance reaches a certain threshold and vote for the other party otherwise. This voting strategy creates a conflict between the coalition parties: it makes one party interested in lowering government performance. The voters accept the prospect of a coalition conflict involving socially costly sabotage activities because it is the 'price' for being able to give any incentives at all. The outcome is the best the voters can secure: they receive a positive expected payoff, the magnitude of which is about 1/4 of the payoff under a 'normal' coalition.

Real-life examples of 'unity government' (or, as it is known in other countries, the 'Grand Coalition'), are coalitions including the main centre-left and centre-right forces in parliament. Episodes of unity government occurred in the recent political history of Austria, Germany, Israel and Italy. It is possible to derive some modest normative implications of the formal analysis, choosing the payoff of the voters as basis of the assessment. The argument is based on the observation that 'unity government' tends to emerge in proportional representation (PR) systems in the presence of 'extreme' parties who make it impossible for either the 'left' or the 'right' to achieve majority. Unity government, in turn, has an inherent accountability problem that is not present in the case of other types of coalition governments. There is thus a benefit to be derived from the moderation and accommodation of an extreme movement in a PR system. While the analysis sheds light on the benefit, the potential costs of trusting extreme movements with government responsibilities are also apparent. It is not possible, however, to estimate the 'cost side' in general. How dangerous it is to include an extreme party into a coalition government instead of forming a Grand Coalition (as it happened in Austria or Italy) depends on the political context and therefore has to be treated as exogenous for this analysis. Therefore, the thesis stops short of an explicit, context-independent, policy recommendation.

Nevertheless, the analysis may help understand some recent developments in the political history of democracies like Austria and Italy. It can be noted that a successful process of moderation and accommodation happened in the case of Communist parties in many European countries and, less controversially, in the case of Green parties. Such an outcome can not be seen in the case of the extreme right movements of Europe.

Divisive politics and accountability. Chapter 3 analyzes a question that is relevant in both proportional and majoritarian systems. It asks the question what is the effect of ideological polarization on political accountability. Might politicians have an interest in polarizing the electorate? Does polarization hamper political accountability?

Chapter 3 models the heterogeneity in the electorate explicitly. In a political system with an incumbent and an opponent politician, there are partisan voters on both sides while the rest of the voters are independent or 'swing voters.'[8] The idea underlying the analysis is that it is the independent voters in a political system who make political accountability operative. The weight of independent voters and the de-

[8]Voter heterogeneity remains implicit in the analysis of Chapter 2 as it concentrates solely on independent voters. Implicitly, however, the analysis of unity government can be thought of as embedded in just the kind of partisan-versus swing voter context that is described in Chapter 3. This framework also would allow the unity government to be the result of the elections rather than only the point of departure.

gree of political polarization is thus of great importance. We may think of elections as determined by the sum of the two separate dynamics: The first force determines the partisan composition of the electorate. This may reflect how the ideological positions of the competing candidates or parties strike a chord with the electorate. The second force is accountability: if the independent voters are satisfied with the record of the incumbent government, they vote to reelect it. (In an equivalent formulation, we may assume that some voters vote on ideological issues – but parties are somewhat uncertain where on the political spectrum the median voter is located in a given year – while some voters vote on a general interest issue like the macroeconomy or the efficiency of government or corruption.) Looking at the interaction of these two forces, the larger then group of independent voters, the more likely it is that they decide the election rather than an ideology shock affecting the partisan composition of the electorate.

Political accountability, as was described above, is a 'contract' between the independent voters and the incumbent politician. Voters reelect the politician after good outcomes and not after bad outcomes. By the prospect of reelection, voters give incentives to politicians to work hard, to refrain from corruption and other forms of rent-seeking, and to invest in the efficient functioning of the state. The higher the probability that the independent voters decide the election the more the politician values their support. For this reason, the more weight independent voters have, the more they can demand from the politician in exchange for their support. In economic terms, by asking for high performance for reelection, they can extract the full expected rent the politician earns by their support.

This insight leads to the main results of the analysis. The incumbent politician may have an incentive to polarize an electorate (i.e. engage in divisive politics) to weaken the forces of political accountability. If such an action forces some independent voters to take sides, that is, become partisan voters of either politician, independent voters lose leverage over the election process. The incumbent needs to put forward a lower effort to win over the independent voters. These voters thus fare worse while the incumbent fares better. More surprisingly, perhaps, the opponent politician also benefits from divisive politics. This is because their probability of winning the election is higher if the electorate is more polarized and the election is more often decided by idiosynchratic shocks rather than by the independent voters. Thus, if the opponent politician has the possibility to engage in divisive politics, they have the incentive to do so.

1.3 The study of fiscal competition

The study of fiscal competition is the study of how the fact that tax bases may relocate between jurisdictions affects the policy of these jurisdictions. The mobile tax bases in question may be people, firms, factories, or capital in general, among other things. The jurisdictions may be municipalities, regions, member states in a federation, countries, or whole economic unions. The relevant policy (i.e. the instrument of competition) is in most studies a tax rate or a set of tax rates (thus the concept of 'tax competition'), but the composition of public expenditures may be an instrument as well. Before turning to the contribution of the thesis to the study of fiscal competition, this section surveys some aspects of the development the field.

1.3.1 History and previous results

The notion that the mobility of certain tax bases may impose constraints on the tax and expenditure policies of jurisdictions and lead to inefficiencies was formulated by Oates (1972; Ch. 4). It provided a counter-argument to the view of Tiebout (1956), who argued that decentralized provision of local public goods would lead to a beneficial competition where citizens can move to a jurisdiction where the public policies (and taxes) correspond to their tastes. The same notion received a formal treatment somewhat later by Wilson (1986), Zodrow and Mieszkowski (1986) and Wildasin (1988).

There are several aspects of the general development of the field that are worth pointing out. First, the study of fiscal competition emerged in a classical ('Pigouvian') Public Finance context dealing with the provision of local public goods – but experienced a 'game-theoretic turn' from the late 1980's. The early literature built on the classic analysis of public goods by Samuelson (1954) and Musgrave (1959). The 'game theoretic turn', while not rendering the classical theory of public goods irrelevant, seems to be a natural development since fiscal competition makes policy making inherently strategic: the optimal policy of one jurisdiction depends on the policy of the others.

Second, while the theory of fiscal competition was first spelled out in the context of local public finance – with researchers being interested in how U.S. municipalities were competing for residents and businesses – a central policy debate of the last two decades redirected much of the literature to concentrate on the international competition for investment capital.[9] In the context of the local public finance literature it was natural to consider the local property tax as the sole tax instrument and expenditures on schooling as the sole public good provided by the municipalities. The more recent literature most often concentrates on the corporate income tax as the instrument of fiscal competition between states.

Third, the debate between the benevolent-state view of classical Public Finance and a more skeptical view of the state in the Public Choice school left a mark on the literature on tax competition as well. 'Leviathan models' of tax competition (first put forward by Brennan and Buchanan (1980)) postulate that the state acts to maximize tax revenue instead of public welfare, implying that a portion of public spending will be wasteful.[10] Tax competition among 'Leviathans' may be efficiency enhancing as it restricts wasteful spending. Since the opposing views differ only in the evaluation, rather than the supposed mechanism, of tax competition it is difficult to reduce the difference to an empirically

[9]See, e.g. Wildasin (2006) for an overview of the field's history in this spirit.

[10]See also analyses by Edwards and Keen (1996) and Rauscher (1998) on the Leviathan hypothesis. Mechanisms through which tax competition may be welfare-enhancing have been reviewed by Wilson (1999) and Wilson and Wildasin (2004).

meaningful question. Therefore, for the literature of tax competition at least, this debate remains a philosophical one.

Recent surveys of the literature include Fuest et al. (2005) who concentrate on international tax competition for mobile capital; the mostly verbal overviews by Wildasin (2006) and Wilson (1999); the monographs by Haufler (2001) and Wellisch (2000); and a survey on the empirical studies (with an emphasis on local public finance) by Brueckner (2003).

Theoretical insights. A fundamental insight of the literature is that fiscal competition (the competition of jurisdictions for mobile tax bases) may lead to the reduction of public revenues and to the underprovision of public goods. The underprovision result has been shown to be fairly robust to modifications of the basic setup; at least under the realistic assumption that only distortive taxation is available. It has been recognized early in the literature that jurisdictions may compete through more than a single tax instrument. Bucovetsky and Wilson (1991) analyze fiscal competition with two tax instruments: a tax on capital and a tax on labor. They find that tax competition may change the tax structure as jurisdictions shift taxation from the mobile to the less mobile resource. But the authors also obtain the underprovision result. In another influential paper considering multiple fiscal instruments Keen and Marchand (1997) analyze a situation where, beside a public good, jurisdictions also provide a public input that improves the productivity of capital. In this case, there is an unambiguous underprovision of the public good, and, at the same time, an 'overprovision' of the public input, at least in a relative sense (compared to the public good).

There is thus an overall sense in the literature that uncoordinated fiscal competition might lead to an undertaxation of mobile tax bases, an underprovision of public goods, and that tax coordination among jurisdictions might therefore lead to a welfare improvement for all. There are some considerations, however, that might contradict this view; either because they point at a force that works against the underprovision result or because they point at a factor that might make policy coordination countereffective (i.e. not beneficial for all). In the following, three such considerations will be presented. These approaches are relevant for the thesis, because Chapter 4 also considers a framework where attempts at tax harmonization can be countereffective, reducing welfare in all countries.

First, it has been recognized early that differences in the size of jurisdictions (cities, regions, countries) might stand in the way of policy coordination. Bucovetsky (1991) shows that small jurisdictions might, in equilibrium, set a lower tax rate than larger ones.[11] The reasoning is the following: Capital flows equalize the rate of return to capital between jurisdictions. The rate of return on capital, in turn, depends on the capital stock *per capita*. Thus, a given capital flow affects a larger jurisdiction less than a smaller one – in a per capita sense. The larger jurisdiction perceives its tax base as less elastic and sets, in equilibrium, a tax rate that is higher than the one in the small jurisdiction (although still lower than the optimum in autarky). By setting a relatively high tax rate the

[11]Kanbur and Keen (1993) reach further results on asymmetric tax competition in the context of cross-border shopping. An intuitive overview is provided by Haufler (2001).

large jurisdiction provides a positive externality for the small one: Residents of the small jurisdiction are better off than their counterparts in the large jurisdiction. It might be the case that a perfect policy coordination ('merging' the jurisdictions or coordinating on an equivalent – and uniform – tax rate) makes the small jurisdiction worse off.

Second, it has been shown that 'agglomeration forces' change the nature of tax competition: most importantly, the 'race to the bottom' might not occur. Baldwin and Krugman (2004) introduced agglomeration effects, an insight from the so-called new economic geography, to the study of tax competition.[12] The presence of agglomeration forces mean that there are external effects involved in certain economic activities, which affect location choices. Simply put: certain industries (or even industry as such) tend to concentrate in certain small geographic areas due to external effects. If such effects are at work, a mild increase in the tax rate of the 'core region' will not have the effect of driving out some firms to the 'periphery.' A large increase, however, might force the whole industry to relocate. Thus, in the analysis of Baldwin and Krugman (2004) the core region will set a high tax rate; as high as it can without risking it's role as the industrial 'core.' As a consequence, the 'periphery region' is not able to compete for the global capital stock and thus sets its tax rate without consideration of outside forces. Tax competition in the presence of agglomeration forces is thus one-sided and very limited. The authors also analyze the possibilities of harmo-

[12] A similar approach was taken by contemporaneous papers by Andersson and Forslid (2003), Kind et al. (2000) and Ludema and Wooton (2000). See also the survey in Baldwin et al. (2003).

nization policies. They show, among others results, that perfect harmonization might actually be harmful for all countries.

Last, a somewhat separate branch of the literature pointed out that certain constitutional arrangements in multi-level government might result in the overtaxation of some tax bases. The theory of 'vertical tax externalities' in fiscal federalism concentrates on cases where the same tax base is taxed by both the 'state' and the 'federal' level.[13] In cases of 'concurrent taxation' by several levels of government the tax base externality goes in the opposite direction than in 'horizontal' tax competition: if the federal level raises its tax rate, the state sees its tax base diminished and vice versa. Thus a vertical tax externality results in too high tax rates on the shared tax base. Whether this effect is stronger than the downward pressure on taxation from 'horizontal' tax competition is, a priori, not clear.[14]

Empirical results. There has been a clear reduction in the statutory corporate tax rates of developed countries since the beginning of the 1980s. While the theory of international competition for mobile capital provides a plausible and intuitive explanation, there are several factors that make it a difficult task to identify the theoretical effect in real world data. The main factor is the complex relationship between Cause and Effect – the tax code and the incentive for business to invest in a given country. The tax code affects the incentives to invest in many

[13]Early studies are Cassing and Hillman, 1982; and Flowers, 1988. An overview of the literature and the questions involved is provided by Keen (1998).

[14]See, e.g. Wrede (1996) and Keen and Kotsogiannis (2002).

ways other than the statutory tax rate. The 1980s and 1990s saw a series of tax reforms (in the USA, UK and Germany among other countries) that reduced the statutory tax rate while broadening the tax base. How broad the definition of the tax base is, depends on whether certain economic activities enjoy exceptions from the general tax rules and on the existence and extent of tax allowances.

An important contribution to the conceptual exploration and measurement of the relevant variables is by Devereux, Griffith and Klemm (2002). The authors emphasize that countries may compete with eachother for investment on several 'margins.' The relevant tax rate for the firms' decision about an additional marginal unit of investment is not the statutory tax rate but the *effective marginal tax rate*. But many investment decisions are made on a different margin. For discrete (or 'lumpy') investment choices (e.g. the decision of a multinational firm to locate a whole factory) the relevant tax measure is the proportion of total profit taken in tax or, in other words, the *effective average tax rate*. Based on another branch of the literature on the behavior of multinational corporations (see e.g. Haufler and Schjelderup (2000)) the authors note that even the *statutory tax rate* can be a relevant measure of tax competition: on this 'margin' countries compete not for real economic activities but for flows of taxable profit in the course of multinational 'profit shifting.' All of these three mechanisms could be reasonably reflected by the theory of tax competition. But considering policy recommendations it is important to understand which one is empirically more important than the others.

Devereux, Griffith and Klemm establish the following stylized facts for a group of 18 developed countries: the statutory corporate tax rates have clearly fallen over the 1980s and 1990s (although tax revenues on corporate income have remained broadly stable as a proportion of GDP since 1965); the effective *marginal* tax rate has remained stable over the 1980s and 1990s; finally, the effective *average* tax rates for projects earning positive economic profits have fallen over the 1980s and 1990s. The authors conclude that the stylized facts are consistent with a theory of tax competition for large (discrete) investments but also with an explanation based on the importance of multinational 'income shifting'.

Building on these insights Devereux, Lockwood and Redoano (2008) estimate how countries react to changes of corporate tax rates in other countries. They find that a 1 percentage point reduction in the average statutory tax rate in other countries reduces the tax rate in a given country by almost 0.7 percentage points. The authors find somewhat weaker evidence for tax competition in effective marginal tax rates, but find no statistically significant 'cross-tax' effects (foreign statutory tax rates affecting home effective marginal tax rate and vice versa).[15]

The empirical work on tax competition on the local level is somewhat more extended than on the international level. Early studies include Case, Rosen and Hines (1993) who concentrate on government expenditure by US states and Ladd (1992) who concentrates on taxation by US counties. Explicit tax reaction functions have been estimated, among others, by Heydels and Vuchelen (1998) and Brueckner

[15]A similar research question was analyzed in the unpublished paper of Altshuler and Goodspeed (2002).

and Saavedra (2001) concentrating on property taxation in Belgian and US municipalities, respectively. Brueckner (2003) surveys the literature.

This literature is closely related to the study of yardstick competition reviewed above. But while the 'reaction function' of jurisdictions has the same structure in both cases, the reason for the policy interdependence is different. Local decision makers might react to the decisions of their neighbors because of the mobility of tax bases (fiscal competition) or because their constituents evaluate home policies as compared to the policies in neighboring jurisdictions (yardstick competition).

1.3.2 Contribution of the thesis

Minimum taxes in repeated tax competition. The original literature on tax competition arrived naturally at the conclusion that tax harmonization (with instruments like perfect harmonization or a lower bound on admissible tax rates) is beneficial for all jurisdictions in preventing a race to the bottom. This conclusion corresponded so closely with the intuition of the scholars in the field that one is tempted to view it more as a premise than a conclusion.[16]

The conclusions of the early tax competition literature were reflected in policy recommendations. Most prominently, the Ruding Committee (Commission of the European Communities, 1992) recommended

[16]A dissenting opinion was presented based on the Leviathan-view of the state, stating that the state bureaucracy has a vested interest in its own growth which ultimately is harmful to general welfare. But this view differed from the original one in the normative evaluation of tax competition rather than in the description of its functioning and underlying mechanism.

the introduction of a lower bound on admissible corporate tax rates (a 'minimum tax') in the European context.

The question asked in Chapter 4 is whether the recommendation for a minimum tax is still supported if tax competition is viewed as repeated rather than static ('one-shot') interaction.[17] The question is relevant because states are long-lived entities and tax competition plays out in real time as a dynamic process: each time a state changes its tax rule other states can react in response.

Analyzing dynamic strategies in tax competition between states means analyzing 'collusion' between states in a similar way as oligopoly theory analyzes collusion between firms. While the parallel might sound unrealistic, there are reasons why it might be fitting. First, collusion between states (or jurisdictions in general) is, as opposed to that between firms, legal. Second, states do cooperate on many issues continuously, both related and unrelated to tax policy. The existing cooperations and interactions in other fields, while not taken into account explicitly in the analysis, might reinforce the potential of dynamic strategies to sustain cooperation in tax policy.

In the analysis of Chapter 4, repeated interaction allows states to employ simple dynamic strategies in order to sustain a high tax rate

[17]Chapter 4 follows a small but growing literature in addressing repeated tax competition. As described in more detail in Section 4.2, different aspects of repeated tax competition have been analyzed by Coates (1993), Kessing et al. (2006) and, most related to the present work, Cardarelli et al. (2002). None of these studies, however, analyzed explicitly the effect of a lower bound of admissible tax rates.

in all states. In particular, a 'trigger strategy' in the spirit of Friedman (1971) prescribes states to cooperate at a high tax rate in the beginning and keep their tax rates high as long as everyone else does so. In the case of a deviation, all countries are prescribed to revert to the 'race-to-the-bottom' tax rate, the static (one-shot) equilibrium. High tax rates can thus be sustained in a subgame-perfect equilibrium by the prospect of a 'punishment' in case of a deviation. The punishment in turn is credible: if a state expects all others to revert to the race-to-the-bottom tax rate, it can do no better than following suit.

The main finding of the chapter is that a minimum tax may reduce tax rates in all states and lead to a welfare loss. The reason is that it restricts states to impose a harsh punishment in case of a deviation. As the possibility of the punishment becomes less threatening, it is more likely that a state follows the temptation and cooperation collapses.

Chapter 2

Coalitions and political accountability

2.1 Motivation

Do coalition governments suffer from an accountability deficit? When do elections provide the right incentives to coalition governments and when do they fail to do so? Are there situations when reelection incentives induce a conflict among government parties? These questions are addressed in the present paper in a simple model of political accountability.

The accountability deficit of coalition governments is a significant, but often implicit, theoretical hypothesis behind many empirical studies in public finance and political economics. In the study of public debt, many explanations for why coalition governments may run higher budget deficits refer to inefficiencies of coalition decision making. Such explanations include the collective action problem (or 'common pool' problem) in the spending of public funds, the lack of commitment power of coalition partners, and the high number of veto players.[1] These arguments, however, do not take into account the influence of elections on the actions of governments. If there is a high probability that voters remove governments after poor outcomes, coalitions have an incentive to solve the collective action problem.

The paper introduces the possibility of coalition government (a government that consists of more than one decision maker) into the theoretical study of political accountability and analyzes the accountability of coalitions as a problem of team production. Building on analyses by Barro (1973), Ferejohn (1986) and Persson and Tabellini (2000, Ch. 4), it concentrates of the moral-hazard aspect of electoral politics, examining a political economy where voters can give incentives to government with the prospect of reelection: they reelect the incumbent if government 'output' is high enough. To assess the accountability of coalition

[1]These arguments have been put forward in the seminal work by Roubini and Sachs (1989). The robustness of the 'weak government hypothesis', as they proposed it, is disputed by de Haan and Sturm (1997). Volkering and de Haan (2001) find a positive effect of government fragmentation on debt growth and debt in OECD countries. Ashworth et al. (2005) and Solé-Ollé (2006) find supporting evidence for the hypothesis for Flemish and Spanish municipalities, respectively, while the former provide a survey of the literature. Recent studies on the occurrence and success of fiscal adjustments find some, but unstable, effect of coalition governments (see Mierau et al. 2007; Illera and Mulas-Granados 2008).

governments the question is asked: Do voters have to settle for a lower government output if government consists of more than one decision maker?

Two main results emerge from the analysis. First, coalition government in itself does not hamper political accountability. Coalition governments can be given appropriate (collective) incentives as long as, in the presence of an electoral alternative, the coalition as a whole can be voted out of power. The reason is that reelection conditional on government performance works like a *discrete team bonus*, the type of contract between the principal (voters) and the team of agents (coalition government) that was shown by Holmström (1982a) to solve the moral hazard problem in teams.

The second main finding is that the accountability of a coalition government becomes problematic when, in the absence of a real electoral alternative, the government cannot be removed as a whole. In this case coalition parties can not be given appropriate team incentives. To incentivate government performance, voters have to make one of the coalition parties responsible for the outcome. This creates incentives for the other party to reduce government performance (or engage in 'sabotage'). In this way, a conflict emerges between the coalition parties, taking the form of a socially costly contest. The resulting contest between the parties is most closely related to a tournament with 'handicap' or 'head-start advantage' analyzed by Konrad (2002). As a difference to that paper, where the handicap is an exogenous effect related to technology, here it is endogenously determined by the voters' strategy. Accordingly, this paper characterizes the 'optimal handicap' as chosen by the voters. It is shown that voters can secure a positive expected payoff even when facing this type of coalition. It is, however, as low as one-fourth of the payoff that voters can get in the presence of an electoral alternative.

The government form corresponding to this theoretical description is the 'unity government' or, as it is known in some countries, the 'Grand Coalition,' a coalition including the major centre-left and -right parties of a political system. In line with the theoretical analysis, 'unity government' can be defined as a situation where it is certain that (at least) one of the government parties stays in power after the next elections. Such governments played an important role in the recent political history of Austria (1945-66, 1987-2000, 2007-), Germany (1966-69 and 2005-), Israel (1984-90 and 2001-03) and Italy (before 1991). In most of these cases the major centre-left and -right forces formed a coalition because neither bloc achieved a majority in the presence of 'extreme' or 'anti-system' parties.[2]

Since coalition government is a characteristic government form of proportional electoral systems, the analysis has some, decidedly modest, normative implications for the study of proportional representation (PR). According to the analysis the Achilles' heel of PR is that the emergence of 'extreme' parties disrupts the alternation of governments, forcing the formation of a unity government. A political system based on

[2]See Geys et al. (2006) for a study of an extreme party's effect on coalition formation in Belgian local elections. For an overview of the theories and stylized facts of coalition government in the European context see the monography of Laver and Schofield (1990).

PR can thus benefit from the moderation and accommodation of 'extreme' movements: it can preserve the possibility of alternating governments and, with that, government accountability. While the analysis identifies the benefit, the potential costs of trusting 'extreme' movements with government responsibilities are also apparent, even if they may vary from case to case. A successful process of moderation and accommodation happened in the case of Communist parties in many European countries and, less controversially, in the case of Green parties. Such an outcome can not be seen in the case of the extreme right movements of Europe.

The analysis is not meant to decide the question of choice between electoral rules. Nonetheless, it does provide a more satisfying theoretical argument about the possible weakness of coalition governments (and perhaps of PR) than conventional references to the inefficiencies of coalition governments, since it takes into account the role of elections.

2.2 Related literature

Beside the weak government hypothesis discussed above, the present paper is related to three branches of literature in economics and political science. First, it intends to contribute to the literature on political accountability by introducing coalition government into the field of study. The theoretical analysis of political accountability was initiated by the work of Barro (1973) and Ferejohn (1986) as a study of moral hazard in electoral politics.[3] While in most studies the ability of voters to hold politicians accountable is welfare-improving, in recent papers, Maskin and Tirole (2004) and Dewatripont and Seabright (2006) point out potential weaknesses of political accountability. Most related to the present paper is the analysis by Persson, Roland and Tabellini (1997) who, similarly to the present paper, also study the accountability of multiple decision makers (politicians). In their framework, however, both decision makers are accountable separately to the electorate, similarly to the president and the congress in a presidential system. This arrangement allows the authors to analyze the effect of 'checks and balances' in a political system. In our framework, the decision-makers are part of the same elected body, and therefore cannot provide checks and balances against each-other. For an overview of the issues related to political accountability see Persson and Tabellini (2000) and Besley (2006).

Second, the analysis is related to several fields in the theory of incentives. The theory of *moral hazard in teams* is relevant for the accountability of coalitions with electoral alternative. Recent studies, following the seminal work of Holmström (1982a), include Itoh 1991; Che und Yoo 2001; and Battaglini 2006. The theory of *all-pay auctions* become relevant in the case where a conflict emerges between the parties of the unity government. This type of contest was thoroughly analyzed by Hillman and Riley (1989), Hirschleifer and Riley (1992), and Baye, Kovenock and de Vries (1996). As described above, the present paper

[3]Beside the moral-hazard aspect, some analyses introduced an adverse-selection element to the analysis of political accountability (see Banks and Sundaram 1993; Besley and Case 1995a,b; Fearon 1999).

extends work by Konrad (2002) on all-pay auctions with a 'handicap' or 'headstart advantage'. Lazear (1989) provided an early and influential analysis of *sabotage in contests.* More recent analyses include Konrad 2000; Chen 2003; and Münster 2007. Our setup differs from these in that sabotage is not described as a separate (second) instrument of the players, but rather as adverse effort. The terminology is used because this counter-effort hurts government performance.

Last, the paper is related to the literature on retrospective (economic) voting. Developed from an early debate within political science about voter motivation, the modern analysis of economic voting concentrates on the effect of macroeconomic outcomes on the popularity (or vote share) of government.[4] While the literature overwhelmingly supports the hypothesis that governments are held responsible for economic outcomes, recent analyses emphasized the way political institutions influence this relationship. In an influential paper, Powell and Whitten (1993) find evidence for the 'clarity-of-responsibility hypothesis'. According to this hypothesis, governments are punished for bad economic outcomes more severely if the assignment of responsibility for government policy is clearer; that is, in the absence of strong bicameral opposition, in the absence of a strong committee system in the legislature and, particularly relevant in our context, in the case of one-party majority government – as opposed to coalitions. Complementary to this finding, Anderson (1995, Ch. 6) finds in a comparative study of five European

[4]Nannestad and Paldam (1994) and Lewis-Beck and Paldam (2000) provide surveys of the field while the volume edited by Norpoth, Lewis-Beck and Lafay (1991) gives account of many aspects of the literature in more detail.

democracies that economic conditions shift voter support not only between government and opposition, but also *among* coalition partners. By investigating theoretically the possibility of voters to reward or punish coalition governments, the present paper intends to contribute to this literature.

The next section analyzes the accountability of a coalition government in the presence of an electoral alternative. Section 3 turns to the case where political accountability is problematic: the case of unity government. Section 4 concludes.

2.3 Accountability of electoral blocs

Consider an economy with an incumbent government L, an opponent R (whose role, as usual in electoral accountability models, is perfectly passive) and a continuum of identical voters, represented by voter I. The incumbent government L consists of two decision makers (or 'factions'), M and N, thus it will be called an 'electoral bloc'.

Each faction in the incumbent government chooses a (non-negative) effort $e_i \in \Re_+$, $i \in \{M, N\}$, simultaneously and non-cooperatively. Voters can observe only the sum of efforts, $e = e_M + e_N$. Effort is beneficial for voters; their payoff w is given as $w = e$. After e becomes public, elections are held, where the incumbent bloc L is facing an opponent R. Voters are indifferent between the electoral blocs at the election stage.

The electoral bloc winning the election receives a rent of value v. If the incumbent bloc remains in power, the factions share the rent according to exogenously given proportions α_i, $i \in \{M, N\}$ with $\alpha_i \geqslant 0$ and $\sum \alpha_i = 1$. Specifically, the payoff of faction i, $i = \{M, N\}$ is given as

$$u_i = \begin{cases} \alpha_i v - e_i & \text{if } L \text{ reelected} \\ -e_i & \text{else} \end{cases} . \tag{2.1}$$

At the beginning of the game, voters coordinate on a voting strategy. We consider the following class of voter strategies: The representative voter I will vote for electoral bloc L if $e \geqslant \overline{e}$, $\overline{e} \in \Re_+$, otherwise she will vote for bloc R. Thus, a strategy is given by the value of $\overline{e}$. Such a strategy is sometimes referred to as a 'simple retrospective voting rule' (e.g. Persson et al., 1997).

The sequence of events is as follows: (1) The voters announce a voting strategy for reelecting the incumbent electoral bloc. (2) The factions choose their respective efforts e_i, $i \in \{M, N\}$. The sum of efforts e is observed by the voters. (3) Elections take place. The newly elected government earns the rents from office and the game ends.

Since, at the election stage, voters are indifferent between the electoral blocs, it is weakly optimal for them to follow their announced voting strategy, whatever it was. Therefore, it is reasonable to concentrate on subgame-perfect equilibria (SPE), where politicians expect voters to execute their announced voting strategy, and voters indeed do so. This allows us to identify the SPE that are optimal from the point of view of the voters, since their announcement is the first move.[5] In this way, as Persson et al. (1997) point out, we analyze the 'potential' of electoral

[5]Other SPE can be supported by less plausible beliefs on the politicians' side. For instance, there exists a SPE where politicians expect never to be reelected whatever the announced voting strategy was. Thus, they exert no effort. Since

accountability in different institutional settings, that is, in different constellations of coalition politics.

Proposition 2.1 *In equilibrium, voters reelect the incumbent electoral bloc L if and only if* $e \geqslant \overline{e}$, $\overline{e} = v$. *The factions of the incumbent electoral bloc put forward an effort* $e_i = \alpha_i v$ *for* $i = \{M, N\}$, *so that* $e = \overline{e}$.

Proof. We solve the game backwards. Since voters are indifferent between the electoral blocs at the election stage, it is (weakly) optimal for them to execute their announced voting strategy, whatever that is. We can now turn to the effort stage. Taking effort $e_j,\ j \in \{M, N\}$ as given, faction i ($i \in \{M, N\}, i \neq j$) compares two relevant alternatives: exerting just enough effort to satisfy the voters or no effort at all. Satisfying the voters is optimal if $\alpha_i v - (\overline{e} - e_j) \geqslant 0$ which is equivalent to the condition $\alpha_i v \geqslant \overline{e} - e_j$. This expression is an incentive constraint: faction i will not exert more effort than $\alpha_i v$ to gain reelection. The sum of efforts can thus never exceed v in equilibrium. If, however, voters set $\overline{e} = v$, it is an equilibrium that factions set $e_i = \alpha_i v$, $i = \{M, N\}$, since their incentive constraints are just binding. ■

Proposition 1 shows that in a simple political accountability game, voters can extract the full rent from the incumbent government, even if it consists of multiple decision makers (factions). In other words, coalition governments can be held accountable. The result is closely related to Theorem 2 of Holmström (1982a). Elections provide here a particular type of contract (a discreet team bonus) between the voter (principal)

---

voters expect never to see positive effort, they cannot do better than choosing strategy 'never reelect the incumbent'.

and the incumbent factions (agents). This is, however, exactly the type of contract that solves the free-rider problem of teams in the analysis of Holmström.

The result does not depend on the number of factions that constitute the incumbent electoral bloc. What is crucial, however, is the presence of an electoral alternative to the incumbent government. The ability to 'reward' or 'punish' the government as a whole allows the voters to give appropriate team incentives to the incumbent factions.

2.4 Accountability of the unity government

As the analysis of the previous section shows, voters can always provide appropriate collective incentives for the government as long as there is an electoral alternative. Accountability becomes problematic if it is certain that (at least) one of the governing parties stays in power after the elections, a description corresponding to the real-life government form that is often called a 'unity government' or, in other countries, a 'Grand Coalition.' This case is modeled here as a situation without opposition. The only thing voters can do is to choose between the government parties.[6]

---

[6]In the formulation presented here, the unity government cannot be an outcome of the election. This inconsistency can be remedied in a more complex voting game without changing the qualitative results of the analysis. Such a voting game with an additional extreme party and a division of voters to 'partisan' and 'swing' voters, where swing voters decide the elections with some

Two further conditions are necessary for political accountability *not* to work in our framework. The first condition is that the effort of the government parties cannot be disentangled by the voters. They observe only the sum of efforts, that is, only one measure of government performance. This condition would not hold in a framework where voters could assign the responsibility for every issue (or, more generally, every action) to one of the coalition parties in the spirit of Laver and Shepsle (1990). However, there are reasons why inseparability of responsibility is a plausible assumption in our context. The first reason is that, the most prominent policy decisions require agreement among coalition partners to pass legislation. Moreover, many outcomes of interest (like the state of the economy in general or the level of government spending) are influenced by many factors; responsibility for them cannot be assigned to a single policy act or a single agent.

The second condition for accountability not to work is the possibility of 'sabotage', defined here as costly effort reducing government performance. Neither of these conditions represents a departure from the framework presented in the analysis of electoral blocs. Clearly, none of the factions had an incentive to engage in sabotage in that context. Further discussion of the importance of these conditions is provided after the main results.

Consider an economy with two office-motivated parties, L and R, both in government at the beginning of the game. The parties choose effort $e_i \in \Re$, $i \in \{L, R\}$ simultaneously and non-cooperatively. Neg-

probability, was described in an earlier version of this paper and is available on request.

ative effort is possible, but is costly: the cost of effort is equal to $|e_i|$. There is a continuum of identical voters. The voters observe only the sum of efforts, $e = e_L + e_R$. The voters' utility w is given by $w = e$. After e becomes public, elections are held, where voters can choose between the incumbent parties L and R. The representative voter I wants to induce a high effort by the government parties with her voting behavior, and is inherently indifferent between the two parties at the election stage. The party that wins the election receives a rent of value v_i, $i \in \{L, R\}$. Note that in this case, parties may have different valuations of winning (equal valuations will be discussed as a special case). Party i's ($i \in \{L, R\}$) payoff is thus:

$$u_i = \begin{cases} v_i - |e_i| & \text{if } i \text{ elected} \\ -|e_i| & \text{else} \end{cases} . \tag{2.2}$$

The sequence of events is as follows: (1) The voters announce a voting strategy (see below). (2) The parties choose their respective efforts e_i, $i = \{L, R\}$. The sum of efforts e is observed by the voters. (3) Elections take place. The newly elected government earns the rents from office and the game ends.

We consider the following class of simple retrospective voter strategies: Representative voter I will vote for party i, $i \in \{L, R\}$, if and only if $e \geqslant \overline{e}$, $\overline{e} \in \Re$. Otherwise she will vote for party $j \in \{L, R\}$, $j \neq i$. In this way, a strategy is given by a pair $\{i, \overline{e}\}$. Note that the payoff of party j is strictly monotonously decreasing in effort. Beside the fact that effort is costly, the higher the government effort the less probable that party j wins the elections. On the other hand, the payoff of party i has a discrete positive jump in effort (when e_j is kept constant). Thus we can say that voter strategy gives 'positive incentives' to party i and 'negative incentives' to party j. This means that if sabotage (costly negative effort) is possible, party j has an incentive to employ it. Therefore, we expect $e_i \geqslant 0$ and $e_j \leqslant 0$. To avoid confusion about the signs, let us define $s_j \equiv - e_j \geqslant 0$ as the (non-negative) sabotage of party j.

This game is structurally identical to a particular type of first-price all-pay auctions with the two parties as 'bidders.' The 'bids' are the efforts e_i and s_j. The cost of effort cannot be recovered. The representative voter plays the role of an 'auctioneer.' The 'prize' the parties are fighting for is the rent they receive in case of reelection. Party i wins the prize if and only if $e_i \geqslant \overline{e} + s_j$. The last expression entails a departure from the standard all-pay auction. By setting $\overline{e}$, voters can advantage one of the parties and handicap the other. A model with this structure, a class of contests with 'head-start advantage,' has been analyzed by Konrad (2002). As a difference to that paper, the voter's (auctioneer's) problem plays a major role in our setting, as the 'head-start advantage' here is not of technical, but rather of strategic, nature. (For this reason we will also use the term 'handicap' beside 'threshold level of effort' to refer to $\overline{e}$.) The objective of the voters is also unusual. Since the effort of one party benefits the representative voter, but the effort of the other harms her, her objective will be to maximize the expected *difference* of both efforts. Formally, the voters' problem is

$$\max_{\overline{e}} E(e); \text{ with } E(e) = E(e_L + e_R) = E(e_i - s_j). \tag{2.3}$$

To solve the game we apply the equilibrium selection criteria discussed in Section 2. The parties expect voters to execute their announced voting strategy, and voters indeed do so, since it is weakly optimal for them. Each point in the voters' strategy space implements an all-pay auction with a handicap. The voters choose optimally from a restricted set of strategies.

Two points are worth noting about the equilibrium. First, although one party exerts positive effort and the other exerts negative effort (sabotage), the voters do not want to 'punish' the saboteur. Voters know that party behavior is induced by the voters' electoral strategy; that strategy in turn is designed to maximize voters' expected payoff. Ultimately, the emergence of sabotage is the price voters have to pay for being able to give incentives at all; to avoid sabotage, they would have to renounce from any incentive effect (e.g. by not making their voting behavior conditional on government performance). It would also mean that they earn a zero payoff with certainty.

Second, voters can choose which party they make responsible for the government's performance. Choosing optimally, as we will see, voters will give positive incentives to the party with the higher valuation (and choose an appropriate performance threshold). The fact that voters can choose which party to make responsible for the outcome may seem unrealistic in the context of the application. In a real-life example, it could be that the prime minister's party is automatically viewed responsible for the outcome. But in that case the coalition partner will have an incentive for sabotage; and the resulting equilibrium would have the same structure as described here. In summary, both of these aspects might be 'unrealistic' in the context of the real-world political application. But replacing them with more realistic assumptions would not alleviate the accountability problem of the 'unity government'; if anything it would make it worse by restricting the voters' set of possible strategies.

Proposition 2.2 *Let us assume, without loss of generality, that* $v_R \geqslant v_L$. *In equilibrium, voters give their vote to party R if* $e \geqslant \overline{e}$, *and L otherwise; the optimally chosen reservation utility is* $\overline{e} = \max\{\frac{v_R v_L}{v_R + v_L}, v_R - v_L\}$.

Further, parties choose mixed strategies to determine their effort levels:

(i) If $\frac{v_R v_L}{v_R+v_L} > v_R - v_L$, *then the unique mixed-strategy equilibrium of the effort subgame is described by following cumulative distribution functions:*

$$H_L(s_L) = \begin{cases} 0 & \text{for } s_L < 0 \\ \frac{\overline{e}}{v_R} + \frac{s_L}{v_R} & \text{for } 0 \leqslant s_L \leqslant v_R - \overline{e} \\ 1 & \text{for } s_L > v_R - \overline{e} \end{cases} \tag{2.4}$$

$$H_R(e_R) = \begin{cases} 0 & \text{for } e_R < 0 \\ \frac{\overline{e}-(v_R-v_L)}{v_L} & \text{for } 0 \leqslant e_R < \overline{e} \\ [1 - \frac{v_R}{v_L}] + \frac{e_R}{v_L} & \text{for } \overline{e} \leqslant e_R \leqslant v_R \\ 1 & \text{for } e_R > v_R \end{cases} . \tag{2.5}$$

(ii) If $\frac{v_R v_L}{v_R+v_L} \leqslant v_R - v_L$, *then the unique mixed-strategy equilibrium of the effort subgame is described by following cumulative distribution functions:*

$$H_L(s_L) = \begin{cases} 0 & \text{for } s_L < 0 \\ [1 - \frac{v_L}{v_R}] + \frac{s_L}{v_R} & \text{for } 0 \leqslant s_L \leqslant v_L \\ 1 & \text{for } s_L > v_L \end{cases} \tag{2.6}$$

$$H_R(e_R) = \begin{cases} 0 & \text{for } e_R < \overline{e} \\ \frac{e_R}{v_L} - \frac{\overline{e}}{v_L} & \text{for } \overline{e} \leqslant e_R \leqslant v_L + \overline{e} \\ 1 & \text{for } e_R > v_L + \overline{e} \end{cases} . \tag{2.7}$$

Proof. See the Appendix. ■

For the case where $v_R = v_L = v$, the limit of the Proposition can be applied. The voters can choose arbitrarily to which of both parties they give positive incentives. The condition of case *(i)* is fulfilled, thus the optimally chosen threshold level of effort (or handicap) is $\overline{e} = \frac{v}{2}$.

The following example may give an intuition for Proposition 2. Let us assume that $v_R > v_L$. We will solve for the equilibrium party behavior for the case when voters in the first stage chose a retrospective voting strategy described by the pair $\{R, \overline{e}\}$ with $\overline{e} \geqslant v_R - v_L$. Voters in this example choose to give party R positive incentives and vote for it if $e \geqslant \overline{e}$ and for party L otherwise.

We can write the expected payoffs of the parties as follows: $u_R = \Pr(e_R \geqslant \overline{e} + s_L)v_R - e_R$ and $u_L = [1 - Pr(e_R \geqslant \overline{e} + s_L)]v_L - s_L$. As it is established in the analysis of all-pay auctions, this type of game has no equilibrium in pure strategies. We will follow the literature to find the mixed strategy equilibrium.[7]

First, no party will choose a 'bid' (that is, effort or sabotage) that is higher than its valuation, since such a choice gives a negative payoff with certainty. Also, no party will bid below zero, since such a bid is costly and reduces the party's chances of winning compared to bidding zero. Party R thus loses with certainty for any bid $e_R < \overline{e}$, since such a bid loses against the smallest possible bid of the opponent, $s_L = 0$. Therefore, R will not put forward any positive bid below $\overline{e}$. On the

[7]See, for example, Hirschleifer and Riley (1992, Section 10.1.2) and Konrad (2002). The uniqueness of this equilibrium can be shown analogously to the uniqueness proof of Baye et al. (1996).

other hand, party L can secure the prize with a bid of $s_L = v_R - \overline{e}$, earning a secure payoff of $v_L - v_R + \overline{e}$. (This payoff is positive, since we are considering the case where $\overline{e} \geqslant v_R - v_L$.) Thus we expect that R will randomize on the interval $[\overline{e}, v_R]$, earning zero in expectation and that L will randomize over $[0, v_R - \overline{e}]$ earning an expected payoff of $v_L - v_R + \overline{e}$. In this case, the 'handicap' $\overline{e}$ is large enough to turn around the 'ranking' of the players, letting the originally 'stronger' player become the 'weaker' one.

As all actions that a player randomizes over have to give the same expected payoff, we can reach the following equation for party L's actions s_L.

$$H_R(s_L + \overline{e})v_L - s_L = v_L - v_R + \overline{e} \tag{2.8}$$

From this we can solve for $H_R(e_R)$, the cumulative distribution function (c.d.f.) of R's bids. Since the previous equation has to hold for any $s_L \in [0, v_R - \overline{e}]$, the following equation has to hold for any $e_R = s_L + \overline{e}$, $e_R \in [\overline{e}, v_R]$:

$$H_R(e_R) = [1 - \frac{v_R}{v_L}] + \frac{e_R}{v_L}. \tag{2.9}$$

Note that R's bid distribution function has a mass point on zero (since we know he will not bid between zero and $\overline{e}$). Similarly, the equation that describes the expected payoff of R's actions, $H_L(e_R - \overline{e})v_R - e_R = 0$, helps us find the solution for L's bid distribution function $H_L(s_L)$ for $s_L \in [0, v_R - \overline{e}]$.

$$H_L(s_L) = \frac{\overline{e}}{v_R} + \frac{s_L}{v_R} \tag{2.10}$$

Thus L's bid function also has a mass point on zero. The bid distributions constitute an equilibrium, since they were constructed so. Further, the distribution functions allow us to calculate the expected payoff of the voters in equilibrium.

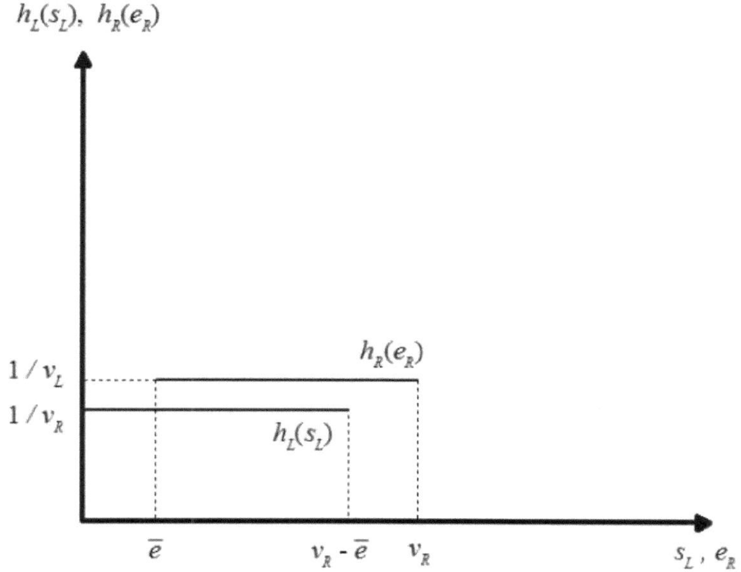

Figure 1. Density functions of party effort in mixed-strategy equilibrium. (Mass points on zero not displayed.)

Figure 1, depicting the *density functions* of party effort and sabotage, gives an intuition about the trade-off the voters are facing. A higher handicap $\overline{e}$ reduces R's expected effort because it raises the threshold below which R does not bid. In effect, the lowest bids of R are turned to zero-bids and the rest is unchanged. At the same time, a higher handicap $\overline{e}$ reduces L's expected sabotage as well; but it is the highest-sabotage bids that are removed. Starting from $\overline{e} = v_R - v_L$ as a reference point, a small increase in the reservation utility reduces expected sabotage more than expected effort, as long as $v_R - v_L$ is small enough. Thus, in that case, the optimal handicap is expected to more than compensate for the difference in valuations between the parties. In other words, if the party with the higher valuation is given positive incentives, it will at the same time be severely handicapped in the voters' optimum.

From Proposition 2 we can calculate voters' payoff as

$$w_I = \begin{cases} \frac{v_R^3}{2v_L(v_R+v_L)} & \text{if } \overline{e} = \frac{v_R v_L}{v_R+v_L} > v_R - v_L \\ \frac{(v_R-v_L)(v_L+2v_R)}{2v_R} & \text{if } \overline{e} = v_R - v_L > \frac{v_R v_L}{v_R+v_L} \end{cases}$$

The first line simplifies to $w_I = \frac{v}{4}$ if $v_R = v_L = v$. Thus, voters get a positive expected payoff even facing a unity government and the possibility of sabotage. Voter payoff is, however, dramatically reduced as compared to the case when an opposition is present. (There, as we saw, voters can receive the full rent v.) This is true because the positive effort of the one party is lower than valuation v with probability 1, while the other party engages in sabotage activity.

It remains to discuss the importance of the two assumptions mentioned above: the possibility of sabotage and the non-observability of individual party effort. It is easy to see that, if sabotage is not possible, voters could give positive incentives to the party with the higher valuation without inducing outright conflict. In that case, again, full rent can be extracted from the party. If, on the other hand, the effort of each party were observed, voters would have no reason to condition their voting strategy on the sum of efforts. Instead, they could induce a 'beneficial' tournament announcing that the party with the higher effort will gain their support.

2.5 Conclusion

This paper analyzed the political accountability of coalitions as a problem of moral hazard in teams. It is shown that a coalition government can be held accountable as long as there is an electoral alternative. Voters can always threaten not to reelect the government, which gives the appropriate team incentives to the government.

The accountability of coalition governments becomes problematic if the voters have no electoral alternative. In this case it is certain that (at least) one of the incumbent parties remain in power after the next elections. Voters cannot give appropriate team incentives to the government, but only choose between the incumbent parties. It was shown that even in this situation, voters can induce a positive expected government performance by making one coalition party responsible for the outcome. This voting strategy creates a conflict among the government parties, making one party interested in reducing government output. The paper solves for the optimal strategy of the voters as 'designers' of the resulting contest between the government parties.

The theoretical description of a coalition government with no electoral alternative corresponds to the real-life examples of 'unity government' or, as it is known in some countries, 'Grand Coalition.' Such coalitions of the main centre-left and centre-right parties typically form in political systems with Proportional Representation (PR) in the presence of extreme parties. Thus, the analysis points at a specific source of accountability deficit in PR systems.

The analysis also provides a counter-argument to the arguments of 'coalition inefficiency' often cited in empirical analyses. Such arguments do not take into account the role of elections. Even if coalitions do face collective action problems, they also have an incentive to overcome them if their reelection probability decreases after inefficient outcomes. The possibility of (no) reelection gives the politicians incentives to act in the citizens' interest.

As an implication for empirical research, it appears that the number of parties included in a government coalition (the variable universally used in empirical work to control for blurred responsibility) may not be the most informative variable. The political constellation in which a (coalition) government operates should play a crucial role. Specifically, it could be useful to identify the weight of extreme parties in the legislature, since this shows ultimately whether there is a real electoral alternative to the government. Information on the electoral margin of government may also indicate the presence of a 'unity government.'

This study did not consider aspects of coalition formation and breakup. The inclusion of this aspect would, however, not affect our results. If, for example, a political faction leaves an electoral bloc, voters have still no incentive to change their strategy in the elections: voting for the incumbent bloc after good outcomes and for the opposition after bad ones. Similarly, an additional coalition formation stage at the beginning of the game would not affect the main driving forces of the analysis: once a governing coalition is in place, the actions of coalition partners are shaped solely by the incentives that voters give them.

A possible path of future research is to analyze the problems of coalition government and political accountability in a more detailed public finance framework. This could allow us to introduce more explanatory factors to analyze the question under what circumstances the 'common pool' effect is (un)likely to emerge in coalition governments.

2.6 Appendix - Proof of Proposition 2

We assumed, without loss of generality, that $v_R \geqslant v_L$. We can divide the representative voter I's strategy space $i, \overline{e} \in R, L \times \Re$ to six ranges. These differ along two dimensions: 1) whether the voters give positive incentives to the party with the higher valuation (whether $i = R$) or to the party with the lower valuation ($i = L$); and 2) in which of three intervals the reservation utility $\overline{e}$ is chosen. The reason to separate exactly these ranges is that the resulting all-pay auction has a different mixed-strategy equilibrium in each of them. In each strategy range we first characterize equilibrium party behavior for a given $\overline{e}$ and search for the voter's optimal choice of $\overline{e}$ within the given range. Then we will be able to make a global statement about I's optimal strategy.

Range 1: Give positive incentives to the party with the higher valuation and handicap him slightly. Suppose voter I's strategy is given by the pair $\{R, \overline{e}\}$ with $v_R - v_L \geqslant \overline{e} \geqslant 0$. The c.d.f. of the parties' equilibrium bid functions are as follows:[8]

$$H_L(s_L) = [1 - \frac{v_L}{v_R}] + \frac{s_L}{v_R} \text{ for } 0 \leqslant s_L \leqslant v_L \tag{2.11}$$

$$H_R(e_R) = \frac{e_R}{v_L} - \frac{\overline{e}}{v_L} \text{ for } \overline{e} \leqslant e_R \leqslant v_L + \overline{e} \tag{2.12}$$

To check that this constitutes an equilibrium, note that L has a negative payoff for all bids below 0 or above his valuation v_L. For any bid s_L between these values L's payoff is $H_R(s_L + \overline{e})v_L - s_L$ which is equal to zero given the supposed form of H_R. On the other hand, the payoff of R for any bid e_R between $\overline{e}$ and $v_L + \overline{e}$ is $H_L(e_R - \overline{e})v_R - e_R$ which, given the supposed form of H_L, is equal to $v_R - v_L - \overline{e}$. It is easy to see that any bid outside this range gives an inferior payoff. Uniqueness of this equilibrium can be shown along the lines of Baye et al. (1996).

The representative voter, seeking to choose the best $\overline{e}$ in the relevant interval $v_R - v_L \geqslant \overline{e} \geqslant 0$, wants to maximize $e = e_R - s_L$. She notes that her choice does not affect L's optimal strategy, but that a higher $\overline{e}$ translates one-to-one to higher effort e_R (in a stochastic sense). Therefore the voter's best option is to choose the upper limit of this interval, that is, $\overline{e} = v_R - v_L$. Her expected payoff is then

$$E[e_R - s_L] = \frac{v_L^2(v_R - v_L)}{2v_R v_L} + \overline{e} = \frac{(v_R - v_L)(v_L + 2v_R)}{2v_R} > 0. \tag{2.13}$$

Range 2: Give positive incentives to the party with the higher valuation while handicapping him strongly. Suppose voter I's strategy

[8]To save space, we will suppress intervals where the c.d.f. of the bid functions is 0 or 1.

is given by the pair $\{R, \overline{e}\}$ with $\overline{e} \geqslant v_R - v_L \geqslant 0$. Then, the c.d.f. of the parties' equilibrium bid functions (identical to the example in the main text) are given by:

$$H_L(s_L) = \frac{\overline{e}}{v_R} + \frac{s_L}{v_R} \text{ for } 0 \leqslant s_L \leqslant v_R - \overline{e} \tag{2.14}$$

$$H_R(e_R) = \begin{cases} \frac{\overline{e}-(v_R-v_L)}{v_L} & \text{for } 0 \leqslant e_R < \overline{e} \\ [1 - \frac{v_R}{v_L}] + \frac{e_R}{v_L} & \text{for } \overline{e} \leqslant e_R \leqslant v_R \end{cases} \tag{2.15}$$

Following the same steps as in Range 1, it can be shown that this constitutes an equilibrium. Now we can turn to the voters' problem.

$$\max_{\overline{e}} E[e_R - s_L] \;\; s.t. \; \overline{e} \geqslant v_R - v_L \geqslant 0. \tag{2.16}$$

$$E[e_R - s_L] = E[e_R] - E[s_L] = \int_{\overline{e}}^{v_R} e_R h_R(e_R) de_R - \int_0^{v_R - \overline{e}} s_L h_L(s_L) ds_L. \tag{2.17}$$

Here, the first equation comes from the fact that the parties randomize independently from each-other, the second uses the usual definition of the expected value of a continuous variable, where mass points at zero can be suppressed. Note that the bid densities are constant on the relevant intervals. Now it is easy to evaluate the integrals to get

$$E[e_R - s_L] = \frac{v_R^2(v_R - v_L)}{2v_R v_L} + \overline{e} - \frac{v_R + v_L}{2v_R v_L}\overline{e}^2. \tag{2.18}$$

We get the first-order condition by differentiating this last expression by $\overline{e}$ and equating the result with zero. This gives us $\overline{e} = \frac{v_R v_L}{v_R + v_L}$.The second-order condition is clearly fulfilled. But we have to make sure that the optimum lies in the considered range $\overline{e} \geqslant v_R - v_L$. This is the case if $v_R \leqslant \frac{1+\sqrt{5}}{2}v_L$. For $v_R > \frac{1+\sqrt{5}}{2}v_L$ we have a corner solution $\overline{e} = v_R - v_L > \frac{v_R v_L}{v_R+v_L}$. To summarize, in this range, voter I's optimal choice for the 'handicap' $\overline{e}$ is given by

$$\overline{e} = \max\{\frac{v_R v_L}{v_R + v_L}, v_R - v_L\}. \tag{2.19}$$

Range 3: Give positive incentives to the party with the higher valuation and give him a head-start advantage. Suppose voter I's strategy is given by the pair $\{R, \overline{e}\}$ with $v_R - v_L \geqslant 0 \geqslant \overline{e}$. The equilibrium bid functions are

$$H_L(s_L) = \begin{cases} \frac{v_R - v_L - \overline{e}}{v_R} & \text{for } 0 \leqslant s_L < -\overline{e} \\ [1 - \frac{v_L}{v_R}] + \frac{s_L}{v_R} & \text{for } -\overline{e} \leqslant s_L \leqslant v_L \end{cases} \tag{2.20}$$

$$H_R(e_R) = \frac{e_R}{v_L} - \frac{\overline{e}}{v_L} \text{ for } 0 \leqslant e_R < v_L - \overline{e} \tag{2.21}$$

The objective function of the voter is

$$E[e_R - s_L] = \frac{v_L^2(v_R - v_L)}{2v_R v_L} + \overline{e} + \frac{v_R + v_L}{2v_R v_L}\overline{e}^2. \tag{2.22}$$

This expression describes a convex parabola. On the one end of the relevant range, at $\overline{e} = 0$, the pay-off is positive, then it decreases below zero as $\overline{e}$ decreases only to start to rise again, reaching zero at $\overline{e} = -v_L$. For higher performance thresholds voter payoff is constant zero, since the outcome of the game is trivial: no party exerts effort as L cannot win. Thus, the optimal 'head-start advantage' in this range is $\overline{e} = 0$.

Range 4: Give positive incentives to the party with the lower valuation and give him a slight head-start advantage. Suppose voter I's strategy is given by the pair $\{L, \overline{e}\}$ with $v_L - v_R \leqslant \overline{e} \leqslant 0$. Then, the equilibrium bid functions resemble those found in Range 1.[9] Also, just as in Range 1, party L's optimal strategy is not affected by the choice of handicap. Voter payoff is

$$E[e_L - s_R] = \frac{v_L^2(v_L - v_R)}{2v_R v_L} + \overline{e}. \tag{2.23}$$

This expression is monotone increasing in $\overline{e}$, so the optimal choice is given by the upper corner $\overline{e} = 0$. Note that voter payoff is negative at this point.

Range 5: Give positive incentives to the party with the lower valuation and give him a significant head-start advantage. Suppose voter I's strategy is given by the pair $\{L, \overline{e}\}$ with $\overline{e} \leqslant v_L - v_R \leqslant 0$. The equilibrium bid functions resemble those found in Range 2. The objective function of the voters is

$$E[e_L - s_R] = \frac{v_R^2(v_L - v_R)}{2v_R v_L} + \overline{e} + \frac{v_R + v_L}{2v_R v_L}\overline{e}^2. \tag{2.24}$$

This is another case where the objective is a convex parabola. Possibilities for the optimum are $\overline{e} = v_L - v_R$ and $\overline{e} = -v_R$. Calculating the payoff for $\overline{e} = v_L - v_R$, we find a negative payoff

$$E[e_L - s_R] = \frac{(v_L - v_R)(v_L + 2v_R)}{2v_R} < 0. \tag{2.25}$$

At the same time, $\overline{e} = -v_R$ (and any choice below that) implements a trivial auction where bids equal zero and L always wins. This option delivers zero payoff to I and is therefore optimal within this range.

[9] Read $-\overline{e}$ for $\overline{e}$; exchange s_i to e_i and vice versa.

Range 6: Give positive incentives to the party with the lower valuation and handicap him. Suppose voter I's strategy is given by the pair $\{L, \overline{e}\}$ with $v_L - v_R \leqslant 0 \leqslant \overline{e}$. The bid functions resemble those in found in Range 3. The expected payoff of the voters is

$$E[e_L - s_R] = \frac{v_L^2(v_L - v_R)}{2v_R v_L} + \overline{e} - \frac{v_R + v_L}{2v_R v_L}\overline{e}^2. \tag{2.26}$$

Note that this objective is identical to the one found in Range 2 up to the constant. (Note also that the constant here is negative while it is positive for Range 2.) Therefore, the optimal handicap is the same as there, $\overline{e} = \frac{v_R v_L}{v_R + v_L}$ (here unconstrained). The voter's payoff is compared below.

The global optimum. After calculating the optimum in each of these ranges we are to rank these (restricted) optima. We will show that the Range-2-optimum represents a global optimum, which proves the Proposition.

First note that Ranges 4 and 5 cannot produce a positive payoff to I, they are thus strictly inferior to Range 1.

Second, note that all strategy ranges are defined such as to include interval limits. This is useful because Range 3 has a corner solution which is an available option in adjacent Range 1. Similarly, Range 1 exhibits a corner solution which is an available option in adjacent Range 2. Therefore, the optimal strategy in Range 2 represents the optimum over Ranges 1, 2, and 3, 4 and 5.

The last step is to show that the optimal strategy in Range 2 is superior to that in Range 6. Here we have to distinguish to cases.

Case A: $v_L < v_R \leqslant \frac{1+\sqrt{5}}{2} v_L$. Here the optimum in Range 2 is $\overline{e} = \frac{v_R v_L}{v_R + v_L}$ just as in Range 6. We have noted that voter I's objective differs across the two Ranges only by a constant. Evaluated at the same reservation utility $\overline{e}$, voter I's payoff is strictly higher in Range 2.

Case B: $v_R > \frac{1+\sqrt{5}}{2} v_L$, or, equivalently, $v_R - v_L > \frac{v_R v_L}{v_R + v_L}$. Here, in Range 2 we have a corner solution at $\overline{e} = v_R - v_L$ giving a payoff of $\frac{(v_R - v_L)(v_L + 2v_R)}{2v_R}$ to I. We can express I's payoff at the optimum in Range 6 as

$$\frac{v_L^3}{2v_R(v_R + v_L)} = \frac{v_R v_L}{v_R + v_L} \frac{v_L^2}{2v_R^2} < (v_R - v_L) \frac{v_L^2}{2v_R^2}, \tag{2.27}$$

where we used Case B's defining inequality. To show that the optimum in Range 6 is inferior to the one of Range 2, we need

$$\frac{(v_R - v_L) v_L^2}{2v_R^2} < \frac{(v_R - v_L)(v_L + 2v_R)}{2v_R}, \tag{2.28}$$

which is fulfilled as

$$\frac{v_L^2}{2v_R^2} < \frac{1}{2} < 1 < \frac{(v_L + 2v_R)}{2v_R}. \tag{2.29}$$

This completes the proof.

Chapter 3

Divisive politics and accountability

3.1 Motivation

Politicians often take ideological positions that, while popular among the initiator's supporters, seem to unite the opposing political camp even more. Such 'divisive politics' leaves the electorate and the party system more polarized, without creating a clear electoral gain for the initiator. In European politics, examples can be found in Spain (in the foreign policy of the People's Party or in the liberal social policies of the Socialist Party) and, more painfully, in the appeal to nationalism and anti-European sentiment in post-transition Poland, Hungary and other countries of Central-Eastern Europe. A common feature of these examples is that the positions taken are distant from the median voter's preferred position and that they help to unite the governing parties as well as the opposition.

This paper demonstrates that politicians may have a strategic incentive to engage in 'divisive politics', even if most of the direct electoral benefit accrues to their opponent. By polarizing the electorate, the incumbent weakens the ability of 'independent' voters to make him accountable for his policies in the common interest. Moreover, the analysis shows that the interests of the incumbent and the opposition may be aligned: the opposition may also benefit from the weakening of political accountability.

The analysis introduces elements of ideology into the analysis of political accountability. The theoretical analysis of political accountability, initiated by the early work of Barro (1973) and Ferejohn (1986), concentrates on the moral-hazard aspect of politics: the conflict of interest between citizens (principals) and politicians (agents). Examples of this conflict of interest are given by corruption and the diversion of public funds by politicians to projects of their preference. Alternatively, one may think of politicians as investing costly effort in the efficient functioning of the state. Voters can make the incumbent act in their interest by offering the reward of reelection in case of good outcomes.[1]

---

[1]Recent developments in the analysis of political accountability include Persson et al. (1997) who study the effect of "checks and balances" in a political system with more than one politician responsible for a policy outcome, and Maskin and Tirole (2004) who point out the potentially negative effects of political accountabilty. For an overview of the issues related to political accountability see Persson and Tabellini (2000) and Besley (2006).

Beside the moral-hazard aspect, some analyses introduced an adverse-selection element to the study of political accountability. Voters in these frameworks would like to choose the more able politicians, beside disciplining the incumbent.[2]

The present analysis is most closely related to Besley (2006, pp. 124-128) who studies a political accountability game with 'partisan' and 'swing' voters. Partisan voters always vote for their preferred party; it is swing voters who excercise political accountability. He finds that electoral accountability is more effective if there is less 'noise' in voters' decisions, if the electorate is less polarized, and if the competition between parties is more even.

The approach taken here differs from that of Besley (2006) in three important respects. First, the present paper models explicitly the politicians' incentives to manipulate the distribution of voters through 'divisive politics'. Second, the incentives of the opponent politician are for the first time analyzed. Third, the present model abstracts from the adverse-selection problem and concentrates on the core moral-hazard aspect of political accountability, as in the model by Persson and Tabellini (2000, Chapter 4).

The main results of this are derived from the analysis of the politicians' incentives to manipulate the distribution of voters prior to the political accountability 'subgame'. Politicians can choose to engage in divisive politics, forcing some swing voters to take sides and become partisan voters of either of the politicians. The result does not merely

[2]See Banks and Sundaram (1993) and Besley and Case (1995). A detailed discussion of such models can be found in Fearon (1999) and Besley (2006).

state that politicians will resort to devisive politics if it brings them an electoral advantage over their opponents (as in Proposition 2 by Besley (2006, p. 127)). The strategic advantage of 'divisive politics' for the incumbent is that swing voters lose some leverage over the reelection, and therefore have to reduce their demands towards the incumbent. 'Divisive politics' pays off for the incumbent even if most of the direct electoral benefit accrues to the opposition. But the analysis also shows that the opponent also gains from divisive politics. As the leverage of swing voters decrease, the probability of the opponent winning the election, in equilibrium, also increases.

3.2 Analysis

3.2.1 The model

Consider an economy with a large number of voters and two politicians, an incumbent A and an opponent B. The incumbent chooses an action $e \in \mathbb{R}^+$, which we call effort. After e becomes public, elections are held where each voter casts a vote for one of the politicians. Either A or B becomes the winner of the elections.

Politicians are office-motivated. The winner of the elections reveives a rent R. The rent from office may be thought of as 'ego rent' but may also be thought of as reflecting the ability to shape (unmodeled) policy. Apart from this rent, the utility of the incumbent depends on the effort he chose before the elections. Effort is costly. We can thus summarize the politicians' expected utility as:

$$Eu_A = \pi R - e \tag{3.1}$$

$$Eu_B = (1 - \pi)R \tag{3.2}$$

where π is the probability that the incumbent gets reelected.

There are three types of voters: partisan voters of either A or B and 'swing voters'. The type of a voter is denoted by $\theta = \{A, 0, B\}$. The utility of each voter increases in the incumbent's effort. Partisan voters receive an additional additive component Ω to their utility if their preferred politician wins the elections. The utility of swing voters does not depend on the identity of the winner; they are inherently indifferent between the politicians. Voter utility can thus be summarized as

$$w_\theta = e + I_\theta \Omega \tag{3.3}$$

for $\theta = \{A, 0, B\}$, where I_θ is an indicator variable that equals 1 if a partisan voter's preferred politician wins the elections and zero otherwise.

The shares of partisan voters are s_A and s_B, respectively. The rest of the voters is independent: their share is s_0 with $s_0 = 1 - s_A - s_B$. The share of partisan voters is stochastic; the uncertainty resolves only at the election stage. The voter shares are $s_A = \overline{s_A} - \varepsilon$ and $s_B = \overline{s_B} + \varepsilon$, where ε is a mean-zero random variable characterized by a continuous c.d.f. $F(\varepsilon) : [-k, k] \to [0, 1]$. (A restriction on the distribution parameter k that ensure the non-negativity of vote shares will be given below after some further definitions.) The nature of the uncertainty and the distribution of the random variable ε are common knowledge.

At the beginning of the game, the incumbent makes a choice $D \in \{0, 1\}$ whether to engage in 'divisive politics.' Divisive politics ($D = 1$)

forces a fraction of swing voters to take sides and turns them into partisan voters. The share of voters turning from swing voters to partisan ones is Δ. A fraction $\lambda \in (0, 1)$ of these voters become partisan voters of the incumbent while the rest $(1 - \lambda)$ becomes partisan to the opponent politician. If the incumbent chooses not to engage in divisive politics, the expected shares of partisan voters are $\overline{s_A} = \overline{s_B} = b$. Divisive politics results in the voter shares $\overline{s_A} = b + \lambda\Delta$ and $\overline{s_B} = b + (1 - \lambda)\Delta$. None of the results below depend on the simplification that both parties initially have an equal share of partisan voters. The substantive assumption is that swing voters are *sometimes* pivotal. Positive vote shares are ensured by $k < b$, and $\Delta < 1 - 2b$. To avoid corner solutions, the analysis concentrates on the case where $b + k > 1/2$. This assumption means that whatever the swing voters do, both politicians have a positive probability of winning.

After the incumbent made this decision, but before he makes the effort choice, the swing voters choose (and announce) a 'simple retrospective voting strategy' for the elections.[3] A simple retrospective voting strategy is sufficiently described by a threshold effort level $\overline{e}$. By announcing $\overline{e}$ the swing voters make the non-binding announcement that they will vote for the incumbent if and only if he chooses an effort higher or equal to $\overline{e}$. This class of strategies enables the swing voters to attain the highest payoff *given* the choice of divisive politics by the incumbent politician. It is crucial for the argument, however, that vot-

[3]Such voting strategies, or as they are sometimes referred to, 'simple retrospective voting rules' are discussed in detail by Persson et al. (1997) and Persson and Tabellini (2000).

ers cannot condition their vote on *whether* the politician chose divisive politics. This assumption tries to capture the idea that divisive politics modifies the conditions of electoral competition but is external to the direct means of electoral competition.

The sequence of the moves is, thus, as follows: (1) The incumbent chooses whether to engage in divisive politics ($D = \{0, 1\}$); (2) The swing voters choose and announce voting strategy $\overline{e}$; (3) The incumbent chooses effort e, which is publicly observed; (4) Each voter casts a vote for either A or B. The election winner emerges and payoffs are realized.

3.2.2 Solving the accountability subgame

We solve the game by backward induction. In the last stage, partisan voters always vote for their preferred politician. Swing voters are inherently indifferent between the politicians; at the election stage it is weakly optimal for them to execute the voting strategy they chose at stage (2).[4]

When choosing the effort level, the incumbent politician compares two relevant alternatives. He gains the votes of swing voters by setting $e = \overline{e}$. Any effort level higher than that causes additional costs with-

[4]Though this is a natural assumption, it is an argument of equilibrium selection. Note that any voting action chosen by the individual swing voters constitutes an equilibrium of the election subgame because none of the large number of voters is pivotal. By assuming that swing voters do not deviate from their announced (and optimally chosen) strategy we, in effect, pick the best equilibrium from the swing voters' point of view (for this argument see also Persson et al., 1997, p. 1171).

out any electoral gain and is therefore strictly dominated. The relevant alternative is to set $e = 0$. Any effort level in the intermediate range $e \in (0, \overline{e})$ is more costly without electoral gain and is therefore dominated by zero effort.

The incumbent maximizes his expected payoff according to the formula

$$\max_e E[u_A(e)] = \pi(e)R - e. \tag{3.4}$$

To be able to compare the relevant expected payoffs, we first calculate the incumbent's probability of reelection conditional on his effort choice. If he sets $e = \overline{e}$, the swing voters will vote for him. The vote share A receives is thus $s_A + s_0$, while B gets a vote share s_B. Using the identity $\overline{s_A} + \overline{s_B} + s_0 = 1$, we can express A's reelection probability as

$$\begin{aligned} \pi(\overline{e}) &= \Pr\left(s_A + s_0 > \frac{1}{2}\right) = \Pr\left(\overline{s_A} - \varepsilon + s_0 > \frac{1}{2}\right) = & (3.5) \\ &= \Pr\left(\varepsilon < 1 - \overline{s_B} - \frac{1}{2}\right) = F\left(\frac{1}{2} - \overline{s_B}\right). & (3.6) \end{aligned}$$

(Note that the continuity of $F(.)$ ensures that ties occur with probability zero. Thus, the tie-breaking rule does not enter into the decision problem of the incumbent.) Turning to the alternative, if A chooses $e = 0$, the swing voters will vote for B. Therefore, the incumbent's reelection probability is

$$\begin{aligned} \pi(0) &= \Pr\left(s_A > \frac{1}{2}\right) = \Pr\left(\overline{s_A} - \varepsilon > \frac{1}{2}\right) = & (3.7) \\ &= \Pr\left(\varepsilon < \overline{s_A} - \frac{1}{2}\right) = F\left(\overline{s_A} - \frac{1}{2}\right). & (3.8) \end{aligned}$$

Comparing the payoffs conditional on the choice of effort, we find that the incumbent will choose $e = \overline{e}$ (rather than $e = 0$) if and only if

$$\overline{e} \leqslant [\pi(\overline{e}) - \pi(0)]R = \left[F\left(\frac{1}{2} - \overline{s_B}\right) - F\left(\overline{s_A} - \frac{1}{2}\right)\right] R. \tag{3.9}$$

Intuitively, this relationship can be understood as an incentive constraint: it does not pay for the incumbent to exert more effort to gain the swing voters' support than than the expected benefit he receives from their support. The expected benefit is the increased probability of reelection times the rent in office. The more probable it is that the incumbent wins the election without the swing voters' support (and the less probable it is that he wins the elections in spite of receiving their votes) the less effort he is ready to put forward.

When, at stage (2), swing voters contemplate to set the effort threshold $\overline{e}$, they must take this incentive constraint into account. Their utility increases with $\overline{e}$ up to the level where the politician is indifferent between choosing $\overline{e}$ and zero effort. If the threshold $\overline{e}$ exceeds that level, the incumbent prefers to choose zero effort and the swing voters' utility falls to zero. Therefore, swing voters will set $\overline{e}$ in a way to make the incumbent's incentive condition bind. In effect, the incumbent receives an expected utility equal to his 'outside option' (zero effort); swing voters are able to extract the full rent difference the incumbent receives by their support. We can summarize the results so far in

Lemma 3.1 *Consider the accountability subgame starting in stage (2). In equilibrium, swing voters set the reelection threshold*

$$\overline{e} = \left[F\left(\frac{1}{2} - \overline{s_B}\right) - F\left(\overline{s_A} - \frac{1}{2}\right)\right] R; \tag{3.10}$$

The incumbent always sets $e = \overline{e}$ and gets reelected with probability $\pi = F\left(\frac{1}{2} - \overline{s_B}\right)$.

3.2.3 Divisive politics in equilibrium

At the first stage of the game, the incumbent politician decides whether to engage in divisive politics. Expecting equilibrium behavior in the subgame starting at stage 2, his expected payoff is:

$$Eu_A = \pi(\overline{e})R - \overline{e} = \pi(\overline{e})R - [\pi(\overline{e}) - \pi(0)]R = \tag{3.11}$$

$$= \pi(0)R = F\left(\overline{s_A} - \frac{1}{2}\right)R = \Pr\left(\varepsilon < \overline{s_A} - \frac{1}{2}\right)R. \tag{3.12}$$

Now we can turn to the question, how this expected payoff is affected by divisive politics. Divisive politics ($D = 1$) increases $\overline{s_A}$, increases $\overline{s_B}$ and reduces s_0. Thus, according to the last expression, it unequivocally increases the expected payoff of the politician for the whole range of possible parameter values $\Delta \in (0, 1 - 2b)$ and $\lambda \in (0, 1)$. We can now state the main result of the analysis.

Proposition 3.1 *In equilibrium, the incumbent politician always chooses to engage in divisive politics ($D = 1$) for all parameter values $\Delta \in (0, 1 - 2b)$ and $\lambda \in (0, 1)$.*

Perhaps surprisingly, the incumbent has an incentive to engage in divisive politics even if it overwhelmingly benefits the opponent (that is, even if λ is very close to zero). To see the intuition of this result, consider the incumbent's expected equilibrium payoff. As was shown, this expected payoff equals the incumbent's 'outside option' at the effort stage, that is, his expected utility after setting $e = 0$. The value of the outside option, however, depends solely on the probability that the incumbent's partisan voters are in absolute majority. All swing and B-partisan voters vote against the incumbent after $e = 0$; any redistribution between these voter groups is inconsequential for A's equilibrium expected payoff. Thus, he will engage in divisive politics even if it benefits the opponent more than himself.

3.2.4 Extension: Divisive politics by the opponent

We have seen that it is in the interest of the incumbent to divide the swing voters. It may be interesting to ask whether the opponent politician B has the opposite interest. To operationalize this, consider a modification of the game analyzed above. In stage (1) of the modified game, the opponent B (instead of the incumbent A) makes a decision $D_B = \{0, 1\}$ whether to engage in divisive politics. If he indeed does choose divisive politics ($D_B = 1$), the expected share of partisan voters become respectively $\overline{s_A} = b + \lambda\Delta$ and $\overline{s_B} = b + (1-\lambda)\Delta$. Otherwise the expected share of partisan voters is $\overline{s_A} = \overline{s_B} = b$. The political accountability subgame (stages (2) to (4)) remains unchanged.

It is left to see under what parameter values B prefers divisive politics. Using the equilibrium of the accountability subgame as analyzed in Subsection 2.2, the payoff of B is

$$Eu_B = (1 - \pi(\overline{e}))R. \tag{3.13}$$

Since $\pi(\overline{e}) = F\left(\frac{1}{2} - \overline{s_B}\right)$ and $\overline{s_B}$ is increased by divisive politics over the full parameter range of $\Delta \in (0, 1 - 2b)$ and $\lambda \in (0, 1)$, we reach the following proposition:

Proposition 3.2 *Consider the modified game where the opponent B can engage in divisive politics. In equilibrium, the opponent always chooses divisive politics ($D_A = 1$) for all parameter values $\Delta \in (0, 1 - 2b)$ and $\lambda \in (0, 1)$.*

This result shows that the interests of the incumbent and the opponent are aligned: both benefit if swing voters are turned into partisan voters, however unbalanced the benefits between the two politicians are. In particular, the opponent benefits even when λ is very close to one. The opponent receives a higher expected payoff because, in equilibrium, divisive politics increases the the probability that the incumbent gets removed from office. Remember that in equilibrium the incumbent sets $e = \bar{e}$ and he receives the votes of the swing as well as his partisan voters. The opponent wins the elections in this case only if his partisan voters are in an absolute majority. This probability is increased even by a very small fraction of independents becoming partisan voters of the opponent.

Relying on the results above, it is possible to make the argument that divisive politics may emerge even under less favorable circumstances. It is a corollary of Propositions 1 and 2 that in a setting where it both politicians must engage in divisive politics for it to become effective and divide swing voters, it is an equilibrium in weakly dominant strategies that both politicians indeed choose divisive politics. In that case, divisive politics is a means of collusion of the politicians against the swing voters.

3.3 Conclusion

The analysis has shown that it may be in the interest of both the incumbent and the opponent politician to use divisive politics. Divisive politics forces some swing voters to take sides and thereby reduces their ability to make the incumbent accountable for his actions in the common interest. In the resulting equilibrium, the opponent also benefits from the weakening of political accountability because his election probability increases eventhough the incumbent satisfies the swing voters' demands.

Chapter 4

Minimum taxes and repeated tax competition

4.1 Motivation

The recommendation for countries to agree on a lower bound to admissible corporate tax rates (a 'minimum tax') has been made repeatedly in recent years, especially in the context of the European Union. As a prominent example, the so-called Ruding Committee (Report of the Committee of Independent Experts on Company Taxation, 1992) proposed setting a minimum corporate tax rate of 30% in the EU. The recommendation for a minimum tax is based on the view that countries, engaged in a competition for mobile resources like capital investment, are forced to lower their tax rates to sub-optimal levels. A minimum tax, in this view, could halt the 'race to the bottom' and thus make all countries better off.

This argument rests on a static theory of tax competition as presented in the first theoretical analyses of the subject.[1] In these models countries finance a public good by raising revenue from a mobile tax base (capital) at source. Departing at the uniform tax rate that maximizes global welfare, an individual country can raise its own tax revenue (and welfare) by reducing its tax rate; attracting a larger share of the global tax base at the expense of other countries. Countries thus face a collective action problem: each profit by individually lowering the tax rate but all suffer after others lowered theirs as well. As a consequence, the Nash equilibrium is not Pareto efficient, and a minimum tax that raises tax rates above the Nash equilibrium is welfare-improving.[2]

But is a minimum tax Pareto improving if tax competition occurs repeatedly rather than as a one-shot interaction? This appears to be a natural question since countries are indeed long-lived, if not immortal, entities. The present paper analyzes tax competition as an infinitely repeated game to address this question.

The main result of this paper is that a minimum tax above the static Nash-equilibrium tax rate may reduce the welfare of all countries. The reason is that repeated interaction allows countries to sustain coopera-

[1]See, e.g., Zodrow and Mieszkowski (1986), Wilson (1986) and Wildasin (1988). A detailed survey is provided by Fuest et al. (2005).

[2]This insight from analyses on symmetric countries generalizes for cases where countries are not very asymmetric. Asymmetry introduces redistributional issues; see, e.g., Bucovetsky, 1991; Kanbur and Keen, 1993.

tion through implicit contracts. Lower bounds on tax rates restrict the ability of countries to punish deviators. This makes cooperation harder to sustain.

4.2 Related literature

The present work is related to three strands of literature. First, the static theory of tax competition, as described above, is understood to imply that a minimum tax cannot be harmful (except, perhaps, at an extremely level).[3] The present paper offers a reassessment of this view.

Second, this paper contributes to the small literature studying repeated tax competition. In an early study in dynamic tax competition, Coates (1993) uses a dynamic setting to introduce long-term effects of capital movements to a model with two tax instruments. Kessing et al. (2006) analyze the effect of vertical tax competition on foreign direct investment, where repeated interaction allows the parties to overcome the hold-up problem. Most related to the present analysis is the work of Cardarelli et al. (2002) who study tax harmonization sustained by implicit contracts. As a difference to the present analysis, none of these studies analyzes the effect of a minimum tax.

Finally, the argument that a minimum tax can be harmful in repeated tax competition has parallels in the study of oligopoly in industrial organization. Known in that context as the 'topsy-turvy principle' (see Shapiro 1989), the observation has been made that market condi-

[3]An instance of harmful minimum taxes has, however, been described by Konrad (2007) in a one-shot setting of Stackelberg structure.

tions making very competitive behavior feasible may actually promote collusion.

4.3 The Analysis

Consider an economy with infinite time horizon with periods $s = 1, 2, ...$ There are N identical countries. In each period each country takes a single action, setting a tax rate on a mobile tax base (capital) at source. The tax rate set by country $i \in \{1, ..., N\}$ in period s is t_i^s, taken from the compact set $T_i \equiv [0, 1]$.

Let the one-period payoff of country i be $V_i(t_1, ..., t_N)$. Countries discount the future by a common discount factor $\beta \in (0, 1)$. The present discounted value of payoffs for country i in period 1 is then

$$PV_i = \sum_{s=1}^{\infty} \beta^s V_i(t_1^s, ..., t_i^s, ..., t_N^s). \tag{4.1}$$

The following assumptions impose some structure on the stage game.[4] Let $V_i(t_1, ..., t_N)$ be twice continuously differentiable and strictly quasi-concave in all tax rates. This implies that the iso-payoff curves are convex to the origin. Also, let $V_i(t_1, ..., t_N)$ be increasing in all t_j with $j \neq i$. The payoff of a country is increasing in the tax rate of the other countries, reflecting one of the main insights of standard tax-competition models, the so-called 'tax base effect': If a country increases its tax rate, leaving the tax rates in other countries un-

[4] A similar 'reduced-form' approach has been taken by Konrad and Schjelderup (1999). The present setup is compatible with the properties of the standard model by Zodrow and Mieszkowski (1986).

changed, some (but not all) of its capital relocates to the other countries. Further, let $\arg\max_{t_i \in [0,1]} V_i(t_1, ..., t_N) \in (0,1)$ be single-valued and increasing in all t_j, $j \neq i$. Thus, reaction functions $t_i(t_{-i})$, where $t_{-i} = (t_1, ..., t_{i-1}, t_{i+1}, ..., t_N)$, are well-defined and tax rates are strategic complements.[5] Below it will be convenient to use the notation $V_i(t_i, t_{-i})$.

Under these assumptions a symmetric Nash equilibrium of the stage game exists, and in what follows it will be assumed to be the unique Nash equilibrium.[6] Let t^N denote the Nash-equilibrium tax rate. Note that the Nash equilibrium does not maximize the countries' joint welfare: since one country's higher tax rate has a positive external effect on all others, a concerted increase of tax rates from t^N would leave all countries better off. (Formally, $\partial V_i(t, ..., t)/\partial t > 0$ for $t = t^N$ because $\partial V_i(.)/\partial t_i = 0$ and $\partial V_i(.)/\partial t_j > 0$, $j \neq i$.)

A jointly welfare-maximizing tax rate $t^C = \arg\max_{t \in [0,1]} V_i(t, ..., t)$ exists by virtue of the boundedness of the range of possible tax rates; and by strict quasiconcavity, it is unique. Hence, it must be that $V_i(t^C, ..., t^C) > V_i(t^N, ..., t^N)$; that $\partial V_i(t, ..., t)/\partial t > 0$ for all $t < t^C$; and therefore $t^C > t^N$. In what follows, t^C will be referred to as the 'cooperative' or 'efficient' tax rate.

---

[5]Strategic complementarity is a common feature of tax competition models; see, e.g., Wildasin (1991), Wilson (1991) and Kanbur and Keen (1993).

[6]Uniqueness is not crucial for the results of this paper, but it simplifies the exposition.

We introduce some more definitions to describe strategies in the repeated game.[7] An action *profile* $(t_1^s, ..., t_N^s)$ describes the actions (tax rates) chosen by all countries in a given period. The set of action profiles is defined as $T \equiv \prod_i T_i$. The set of period s *histories* is given by $H^s \equiv T^s$, where T^s is the s-fold product of T, and the initial history is the null set $T^1 = \{\oslash\}$. A history $h^s \in H^s$ is thus a list of s action profiles, identifying the tax rates chosen by all countries up to period $s-1$. The set of all possible histories is

$$H \equiv \bigcup_{s=1}^{\infty} H^s. \tag{4.2}$$

A *pure strategy* for country i describes what tax rate the country would set after all possible histories; it is thus a mapping from the set of possible histories into the set of pure actions,

$$\sigma_i : H \rightarrow T_i. \tag{4.3}$$

Note that 'Nash forever', the strategy profile in which all countries set the static Nash equilibrium tax rate t^N after all possible histories in all periods $s = 1, 2, ...$, constitutes a subgame-perfect equilibrium of the repeated game. Also, reversion to 'Nash forever', a strategy profile in which all countries set the static Nash equilibrium tax rate t^N in periods $s = s', s' + 1, ...$ if a certain history $h^{s'}$ was reached, constitutes a subgame-perfect equilibrium of the subgame starting with that history.

Based on these observations, we concentrate on trigger strategies first analyzed by Friedman (1971). Such trigger strategies prescribe

[7] The concepts and definitions related to the repeated game are used in a standard way, see Mailath and Samuelson (2006, Ch 2).

countries to set the cooperative tax rate as long no deviation is observed; and set the static Nash-equilibrium tax rate forever after a deviation is observed. Formally, the Friedman-type trigger strategy σ_i^F prescribes country i to set $t_i^1 = t^C$; while for periods $s > 1$:

$$t_i^s = \begin{cases} t^C & \text{if } t_j^\tau = t^C \text{ for all } j \text{ and } \tau = 1, ..., s-1 \\ t^N & \text{else} \end{cases} . \quad (4.4)$$

We set out to examine under what circumstances the efficient tax rate t^C can be supported by the profile of Friedman-type trigger strategies $\sigma^F = (\sigma_1^F, ..., \sigma_N^F)$ as an outcome of a subgame-perfect equilibrium (Proposition 1); and how the results are affected by the introduction of a lower bound on admissible tax rates (Propositions 2 and 3).

Proposition 4.1 *There exists a threshold discount factor $\underline{\beta} \in (0,1)$ such that for all discount factors $\beta > \underline{\beta}$ the profile of trigger strategies $\sigma^F = (\sigma_1^F, ..., \sigma_N^F)$ constitutes a subgame-perfect equilibrium of the infinitely repeated game. In this equilibrium, all countries set the efficient tax rate t^C in every period.*

Proof. Let t^d denote the optimal deviation of country i from cooperation, that is, $t^d = t_i(t^C, ..., t^C)$. Then, $V_i(t^d, t^C)$ denotes the payoff of country i if $t_i = t^d$ and $t_{-i} = (t^C, ..., t^C)$. In any given period, country i finds it optimal not to deviate if the following incentive condition holds:

$$V_i(t^d, t^C) - V_i(t^C, t^C) \leqslant \sum_{s=1}^{\infty} \beta^s [V_i(t^C, t^C) - V_i(t^N, t^N)], \quad (4.5)$$

The left hand side gives the immediate gain of deviation; the right hand side gives the cost in foregone future cooperation. Clearly, as β approaches 1, the right hand side grows without bounds, while the left hand side remains constant. Therefore, there exists a $\underline{\beta} < 1$ for which the condition holds with equality. For all $\beta > \underline{\beta}$ it will hold as strict inequality. ■

The next step is to show that a minimum tax $\underline{t}$ in the interval $(t^N, t^d]$ reduces the sustainability of the efficient tax rate. First note that strategic complementarity implies that this interval is non-empty. Note also that the one-shot Nash equilibrium of the tax competition game with the minimum tax becomes $(\underline{t}, \underline{t})$.

Proposition 4.2 *The introduction of a minimum tax $\underline{t} \in (t^N, t^d]$ restricts the range of discount factors for which the efficient tax rate t^C can be supported by trigger strategies as a subgame-perfect equilibrium outcome in the infinitely repeated game.*

Proof. For a minimum tax $\underline{t} \in (t^N, t^d]$ country i finds it optimal not to deviate from the efficient tax rate if the following incentive condition holds:

$$V_i(t^d, t^C) - V_i(t^C, t^C) \leqslant \sum_{s=1}^{\infty} \beta^s [V_i(t^C, t^C) - V_i(\underline{t}, \underline{t})] \tag{4.6}$$

The only difference to inequality (1) appears in the last term. Since $\underline{t} \in (t^N, t^C)$ it follows that $V_i(\underline{t}, \underline{t}) > V_i(t^N, t^N)$; the right hand side becomes smaller for a given β. Therefore, the incentive condition is now violated for $\underline{\beta}$. There exists $\beta' \in (\underline{\beta}, 1)$ that makes the condition hold with equality. For $\beta \in [\underline{\beta}, \beta')$, in the presence of the minimum tax, it is optimal for any country to deviate from t^C in the first period. Cooperation at t^C can only be sustained for the restricted range of discount factors $[\beta', 1)$. ■

The result has a clear intuition. A minimum tax between the 'punishment' tax rate and the 'temptation' tax rate restricts the punishment for a deviation to be milder while leaving the deviation no less tempting.

A higher minimum tax $\underline{t} > t^d$ affects both the temptation and the punishment, making a the assessment more complex. However, under reasonable assumptions it is possible to show that a minimum tax is harmful even in this range.

Proposition 4.3 *The introduction of a minimum tax $\underline{t} \in (t^d, t^C)$ restricts the range of discount factors for which the efficient tax rate t^C can be supported by trigger strategies as a subgame-perfect equilibrium outcome in the infinitely repeated game if both $V_i(t,t)$ and $V_i(t,t^C)$ are weakly concave in t.*

Proof. Without a minimum tax, cooperation at t^C is sustainable for $\beta \geqslant \underline{\beta}$ (Proposition 1). It has to be shown that countries always have an incentive to deviate from t^C in the infinitely repeated game with discount factor $\underline{\beta}$ and a minimum tax $\underline{t} \in (t^d, t^C)$. Define $A(t) = [V_i(t,t^C) - V_i(t^C,t^C)]$ and $D(t) = \frac{\underline{\beta}}{1-\underline{\beta}}[V_i(t^C,t^C) - V_i(t,t)]$. For a minimum tax $\underline{t} \in [t^d, t^C]$, $A(\underline{t})$ represents the advantage of deviation from cooperation, while $D(\underline{t})$ represents the cost (or disadvantage) of deviation. Proposition 2 established that $A(t^d) > D(t^d)$. At the same time, $A(t^C) = D(t^C) = 0$. Therefore, for any minimum tax $\underline{t} = \alpha t^d + (1-\alpha)t^C$ with $\alpha \in (0,1)$ it holds that:

$$A(\underline{t}) \geqslant \alpha A(t^d) + (1-\alpha)A(t^C) > \alpha D(t^d) + (1-\alpha)D(t^C) \geqslant D(\underline{t}). \quad (4.7)$$

The first inequality follows from the weak convexity of $A(t)$ (implied by the weak concavity of $V_i(t, t^C)$), while the last inequality follows from the weak concavity of $V_i(t, t)$. ■

The weak-concavity assumptions are reasonable. In particular, a specification with quasilinear preferences and quadratic production functions of the tax competition model of Zodrow and Mieszkowski (1986) exhibits *strict* concavity of $V_i(t, t)$ and $V_i(t, t^C)$.

4.4 Conclusion

Viewing tax competition as repeated interaction reverses the common assessment of the desirability of agreements on a lower bound on admissible tax rates (a 'minimum tax'). If tax cooperation is sustained by implicit contracts, a minimum tax may trigger a 'race to the bottom' making all countries worse off. The reason is that a minimum tax restricts countries to punish deviators. Eliminating the worst possible outcomes makes the best ones harder to obtain.

The present analysis is based on powerful and simple dynamic strategies involving 'Nash reversion'. Further research could investigate dynamic strategies that are more severe, and extend the present results to the case where countries threaten deviators with 'optimal punishments' of the type described by Abreu (1986).

Summary in German

Die drei Essays der hier vorgelegten Dissertation behandeln zwei thematische Schwerpunkte: die politische Verantwortlichkeit und den Steuerwettbewerb. Zwei grundlegende Prinzipien, ein 'formales' und ein 'substantielles', einen die Beiträge. Das eine einende Prinzip betrifft also die analytische Form, denn die Instrumente und die Erkenntnisse der Spieltheorie kommen in jedem der Aufsätze zum Tragen. Das andere einende Prinzip prägt die Motivation und den Inhalt der Beiträge, die um die Rolle und die Funktionsweise des Staates kreisen bzw. einzelne Aspekte deren beleuchten.

Die Analyse von politischer Verantwortlichkeit. Im Mittelpunkt der Analyse von politischer Verantwortlichkeit steht die Frage, wie die Aussicht auf Wiederwahl das Verhalten von Politikern im Amt beeinflusst. Bevor der Beitrag dieser Dissertation zur Analyse von politischer Verantwortlichkeit formuliert wird, soll die Theoriebildung in diesem Bereich kurz überblickt werden.

Zur Analyse von politischer Verantwortlichkeit im Kontext der Wahlfor-schung wird gewöhnlich eine Taxonomie von Politikermotivation und Wählermotivation verwendet. Die Theorie politischer Verantwortlichkeit geht von der Vorstellung aus, dass Politiker eine Wiederwahl positiv bewerten und sie für anstrebenswert halten. Diese Annahme schließt selbstverständlich nicht aus, dass es auch andere Motive gibt oder geben kann, etwa der Wunsch, Sachpolitik zu gestalten oder ein politisches Erbe zu hinterlassen. Politiker mögen gerade aus diesen (oder aus anderen) Gründen eine Wiederwahl anstreben. Aber um die Mechanismen von politischer Verantwortlichkeit in Gang zu bringen, reicht als notwendige Bedingung aus, dass Politiker eine Wiederwahl anstreben – die Frage, warum dies letztendlich so ist, kann demnach offen gelassen werden. Was die Wählermotivation angeht, betrachtet die Theorie politischer Verantwortung das Wählen als einen rationalen und retrospektiven Akt.

Kapitel 2 - Koalitionen und politische Verantwortlichkeit. Als theoretischer Hintergrund der empirischen Forschung zu Regierungskoalition-en werden oft Argumente über die Ineffizienz von Entscheidungsprozessen in Koalitionen herangezogen. Indes wurde die Frage nicht beantwortet, welcher grundsätzliche Zusammenhang zwischen der Disziplinierbarkeit der Politik durch Wahl oder Wiederwahl und dem Regieren in Koalitionen besteht. Reduziert auf den vertragstheoretischen Kern des Problems stellen Koalitionsregierungen ein 'Teamprodukt' für den Wähler als Prinzipal her, wobei der 'Vertrag' zwischen Wähler und Koalitionsregierung sehr spezifisch und jedenfalls unvollständig ist. In welchem Ausmaß kann das Sanktionsinstrument der Abwahl oder Wiederwahl die Ineffizienzen der Entscheidungsprozesse in Koalitionen mildern? Welche Anreize übt dieses Instrument auf die Politikentscheidungen innerhalb von Koalitionsregierungen aus? Wann gibt es innerhalb von Koalitionsregierungen konstruktive Wettbewerbswirkungen, wann kommt es zu negativen Effekten, etwa zu Sabotage hinsichtlich des Gesamterfolgs der Regierung?

Das erste Hauptresultat des Kapitels sagt aus, dass die Disziplinierbarkeit einer Koalition unproblematisch ist, wenn eine wahre Wahlalternative vorhanden ist. Die Bedeutung dieses Ergebnisses ist, dass es einen Mechanismus aufweist, durch den die eventuellen Ineffizienzen der Entscheidungsfindung in einer Koalitionsregierung gemildert werden können.

Die Disziplinierbarkeit (also die politische Verantwortlichkeit) wird problematisch in einer Situation, in der sich eine Koalition verschiedener Parteien ergibt, zu der es keine echte mehrheitsfähige Alternative gibt, und die als 'Große Koalition' bezeichnet werden soll. Koalitionen dieser Form sind in der jüngsten Geschichte von Ländern wie Deutschland, Österreich, Israel und Italien wiederholt aufgetreten. Die Besonderheit der Großen Koalition ist, dass mindestens eine der beteiligten Parteien mit Sicherheit nach den nächsten Wahlen weiterregiert. Nur Teile der Regierung können abgewählt werden. Die große Koalition als Einheit kann in dieser Situation von den Wählern nicht *in toto* 'belohnt' oder 'bestraft' werden. Modelliert wird diese Situation als eine Regierungskoalition ohne Opposition, also eine, wo die Regierungsparteien die einzigen Walhalternativen der Wähler darstellen.

Das zweite Hauptergebnis des Kapitels beschreibt die beste Strategie des repräsentativen Wählers in einem stilisierten politischen System, wo die 'Große Koalition' regiert. Es wird gezeigt, dass der Wähler den Regierungspolitikern nur dann Anreize setzen kann, wenn er eine der Regierungsparteien für die Regierungspolitik verantwortlich macht. Dies führt aber zu einem Konflikt zwischen den Regierungsparteien.

Kapitel 3 - Politische Verantwortlichkeit und Polarisierung. Im Kapitel 3 wird die Frage gestellt, wie die ideologische Polarisierung auf die politische Verantwortlichkeit auswirkt. Können Politiker ein Interesse an der Polarisierung der Wählerschaft haben? Behindert eine Politik der Polarisierung die politische Verantwortlichkeit?

Im Kapitel 3 wird die Heterogenität der Wählerschaft konstitutiv für das Modell. In einem politischen System mit einem aktuell regierenden und einem opponierenden Politiker ist ein Teil der Wähler 'Stammwähler' bzw. parteigebunden auf der einen oder der anderen politischen Seite, der Rest der Wähler sind unabhängige bzw. 'Wechselwähler'. In der Analyse kommt den unabhängigen Wählern eine wichtige Rolle zu, denn in jedem politi-schen System sind es gerade sie, die politische Verantwortlichkeit in Kraft setzen. Mit der Aussicht der Wiederwahl können die Wähler dem Politiker Anreize geben, damit er für deren Interessen hart arbeitet, Korruption und anderen Formen der privaten Vorteilsnahme abschwört und stattdessen in das effiziente Funktionieren des Staates investiert. Parteigebundene Wähler können aber nicht glaubhaft ihrer Partei damit drohen, nicht auf sie abzustimmen. Am Wahltag ist es für sie prospektiv immer rational, ihre

Partei zu wählen. Die unabhängigen Wähler sind also die einzigen, die mit ihrem Wahlverhalten die Wohlfahrt aller Wähler beeinflussen können. Das Gewicht der unabhängigen Wähler und der Grad der Polarisierung sind daher von großer Bedeutung.

Politische Verantwortlichkeit kann in diesem Modell als ein 'Vertrag' zwi-schen den unabhängigen Wählern und dem aktuell regierenden Politiker beschrieben werden. Die Wähler würden den Politiker nach einem guten Ausgang (Erfüllung des Vertrags) wiederwählen und nach einem schlechten Ausgang abwählen. Je höher die Wahrscheinlichkeit, dass die unabhängigen Wähler die Wahlen entscheiden, umso höher bewertet der Politiker ihre Unterstützung. Je höher das Gewicht der unabhängigen Wähler in der Wählerschaft ist, umso mehr können sie vom Politiker für ihre Unterstützung verlangen. Ökonomisch formuliert können die unabhängigen Wähler, indem sie eine hohe politische Leistung für die Wiederwahl einfordern, die gesamte erwartete Rente, die dem Politiker durch ihre Unterstützung zugute kommt, für sich beanspruchen.

Diese Einsicht führt unmittelbar zum wichtigsten Ergebnis der Analyse. Der aktuell regierende Politiker kann einen Anreiz haben, die Wählerschaft gezielt zu polarisieren (bzw. eine Politik der Trennung zu betreiben), um jene Kräfte zu schwächen, die eine politische Verantwortlichkeit einfordern. Wenn ein polarisierender Akt einen Teil der unabhängigen Wähler dazu zwingt, 'Partei zu ergreifen', wächst das Gewicht der parteigebundenen Wähler, gleichzeitig verlieren die unabhängigen Wähler an Einfluss auf den Wahlprozess. Folglich braucht der regierende Politiker weniger Aufwand zur Gewinnung der unabhängigen Wähler zu betreiben. Entsprechend werden alle Wähler schlechter gestellt.

Es überrascht jedoch vielleicht mehr, dass auch der opponierende Politiker von der polarisierenden Politik profitiert. Die Wahrscheinlichkeit (im 'Nash-Gleichgewicht'), dass er die Wahl gewinnt, ist nämlich höher, wenn die Wählerschaft polarisierter ist, da in solchen Fällen die Wahlen eher durch unerwartete ideologische Schocks als durch die unabhängigen Wähler ent-schieden werden. Das heißt, hat der Oppositionspolitiker die Möglichkeit zu polarisieren, so hat er gewiss einen Anreiz dazu.

Die Analyse von Steuerwettbewerb unter Staaten. In der Analyse von Steuerwettbewerb gilt es zu untersuchen, wie die Tatsache, dass eine Reallokation der Steuerbasen zwischen politischen Einheiten möglich ist, die Steuerpolitik letzterer beeinflussen kann. Das mobile Steuerbasis kann etwa Arbeitskraft, Unternehmen, Fabrik oder Kapital im Allgemeinen sein. Die betroffenen politischen Einheiten können Kommunen, Regionen, Mitgliedstaaten einer Föderation, Länder oder ganze Wirtschaftsunionen sein. Die relevante Politik (und damit das Instrument des Wettbewerbs) ist in den meisten Studien ein einziger Steuersatz, aber es gibt auch Analysen, die Steuerwettbewerb für mehrere mobile Steuerbasen durch mehrere Steuer-instrumenten betrachten.

Kapitel 4 - Mindeststeuersätze und wiederholter Steuerwettbewerb. Die erste Generation der Steuerwettbewerbsforschung kam zu der eingängigen Schlussfolgerung, dass die Steuerharmonisierung (mit

Instrumenten wie perfekte Harmonisierung oder eine untere Grenze der zulässigen Steuersätze) für alle zuständigen Gebietskörperschaften vorteilhaft ist, nachdem sie einen ruinösen Wettbewerb nach unten ['race to the bottom'] verhindert.

Die Schlussfolgerungen aus dieser frühen Steuerwettbewerb-Literatur sind als Empfehlungen in die Politikberatung übertragen worden. Als Beispiel sei die Empfehlung der Ruding-Komitee an die Kommission der Europä-ischen Gemeinschaft (1992) genannt, eine untere Grenze für zulässige Steuersätze ('Mindeststeuersatz') bei der Unternehmensbesteuerung im europä-ischen Kontext einzuführen.

Im Kapitel 4 wird die Frage untersucht, ob die Empfehlung zur Einführung einer Minimalsteuer auch dann gilt, wenn der Steuerwettbewerb als wiederholte statt statische (einmalige) Interaktion betrachtet wird. Die Frage ist relevant, denn Staaten sind langlebige Entitäten und der Steuerwettbewerb spielt sich in Echtzeit als dynamischer Prozess ab: Jedes Mal, wenn ein Staat seine Steuerregeln ändert, können andere Staaten darauf ihrerseits mit Maßnahmen reagieren.

Die Analyse im Kapitel 4 zeigt, dass wiederholte Interaktion Staaten in die Lage versetzt, einfache dynamische Strategien einzusetzen, um hohe Steuersätze in allen Staaten aufrecht zu erhalten. Genannt seien insbesondere sogenannte 'trigger strategies' nach Friedman (1971). Eine solche Strategie schreibt den Beteiligten vor, zu Beginn bei einem hohen Steuersatz zu kooperieren und die Kooperation so lange aufrecht zu erhalten solange alle dabei bleiben. Im Falle einer Abweichung schreibt die Strategie vor, zu dem statischen Gleichgewichtssteuersatz zurückzukehren, nämlich zu dem Steuersatz, der als Ergebnis eines ruinösen Wettbewerbs nach unten entsteht. Ein hoher Steuersatz kann demnach als teilspielperfektes Gleichgewicht durch die Androhung einer 'Strafe' bei Abweichung aufrechterhalten werden. Die Androhung der Strafe ist in der Tat glaubwürdig: Wenn jeder Staat annimmt, die anderen würden zum Steuersatz des ruinösen Wettbewerbs zurückkehren, dann kann sich nach einer tatsächlichen Abweichung keiner besser stellen, als sich an das schlechte Gleichgewicht zu halten.

Die wichtigste Erkenntnis des Kapitels 4 ist, dass sich die Einführung einer Minimalsteuer reduzierend auf die beobachteten Steuersätze in allen Staaten auswirken und dadurch zu Wohlfahrtsverlusten führen kann. Die Minimalsteuer verhindert nämlich, dass Staaten einen Abweichler drastisch abstrafen. Wenn die Bestrafung an Abschreckungskraft verliert, wird es wahrscheinlicher, dass ein Staat der Versuchung abzuweichen nachgibt, so dass die Kooperation zusammenbricht.

Bibliography

[1] Abreu, D. (1986). Extremal equilibria of oligopolistic supergames. Journal of Economic Theory 39, 191-225.

[2] Alesina, A., G.D. Cohen and N. Roubini (1992). Macroeconomic policy and elections in OECD democracies. Economics and Politics 4(1), 1-30.

[3] Alesina, A., N. Roubini and G. Cohen (1997). Political cycles and the macroeconomy. Cambridge, MA: MIT Press.

[4] Allers, M. and J. Elhorst (2005). Tax Mimicking and Yardstick Competition Among Local Governments in the Netherlands. International Tax and Public Finance 12(4), 493-513.

[5] Altshuler, R. and T. J. Goodspeed (2002). Follow the leader? Evidence on European and US tax competition. Mimeo.

[6] Anderson, C. (1995). Blaming the government: citizens and the economy in five European democracies. Armonk, NY: M. E. Sharp.

[7] Andersson, F. and R. Forslid (2003). Tax competition and economic geography. Journal of Public Economic Theory 5(2), 279-303.

[8] Ashworth, J., B. Geys and B. Heyndels (2005). Government weakness and local public debt development in Flemish municipalities. International Tax and Public Finance 12, 395-422.

[9] Arrow, K. (1951). Social choice and individual values. New York, Wiley.

[10] Austen-Smith, D. and J. Banks (1989). Electoral accountability and incumbency. In P. Ordeshook (Ed.): Models in strategic choice in politics. Ann Arbor: University of Michigan Press.

[11] Baldwin, R. E. and P. Krugman (2004). Agglomeration, integration and tax harmonisation. European Economic Review 48, 1-23.

[12] Baldwin, R. E., R. Forslid, P. Martin, G. Ottaviano and F. Robert-Nicoud (2003). Economic geography and public policy. Princeton, NJ: Princeton University Press.

[13] Banks, J. S. and R. K. Sundaram (1993). Adverse selection and moral hazard in a repeated elections model. In Barnett, Hinich & Schofield (Eds.), Political economy: Institutions, competition and representation. Cambridge University Press.

[14] Barro, R. (1973). The control of politicians: An economic model. Public Choice 14 (1), 19-42.

[15] Battaglini, M. (2006). Joint production in teams. Journal of Economic Theory 130 (1), 138-167.

[16] Baye, M., D. Kovenock and C. de Vries (1996). The all-pay auction with complete information. Economic Theory 8, 291-305.

[17] Besley, T. (2006). Principled Agents? The Political Economy of Good Government. Oxford: Oxford University Press.

[18] Besley, T. and A. Case (1995a). Incumbent behavior: vote-seeking, tax-setting, and yardstick competition. American Economic Review 85 (1), 25-45.

[19] Besley, T. and A. Case (1995b). Does electoral accountability affect economic policy choices? Evidence from gubernatorial term limits. Quarterly Journal of Economics 110(3), 769-98.

[20] Black, D. (1958). The theory of committees and elections. Cambridge University Press.

[21] Bordignon, M., F. Cerniglia and F. Revelli (2003). In search of yardstick competition: a spatial analysis of Italian municipality property tax setting. Journal of Urban Economics 54(2), 199-217.

[22] Brennan, G. and J. M. Buchanan (1980). The power to tax: Analytical foundations of a fiscal constitution. Cambridge University Press.

[23] Brueckner, J. K. (2003). Strategic interaction among governments: An overview of empirical studies. International Regional Science Review 26(2), 175-188.

[24] Brueckner, J. K. and L. A. Saavedra (2001). Do local governments engage in strategic tax competition? National Tax Journal 54, 203-229.

[25] Buchanan, J. M. and G. Tullock (1962). The Calculus of Consent. Ann Arbor: University of Michigan Press.

[26] Bucovetsky, S. (1991). Asymmetric tax competition. Journal of Urban Economics 30, 167-181.

[27] Bucovetsky, S. and J. D. Wilson (1991). Tax competition with two tax instruments. Regional Science and Urban Economics 21, 333-350.

[28] Cardarelli, R., E. Taugourdeau and J.-P. Vidal (2002). A repeated interactions model of tax competition. Journal of Public Economic Theory 4 (1), 19–38.

[29] Case, A.C., H. S. Rosen and J. R. Hines (1993). Budget spillovers and fiscal policy interdependence: Evidence from the states. Journal of Public Economics 52(3), 285-307.

[30] Cassing, J. H. and A. L. Hillman (1982). State-federal resource tax rivalry: The Queensland Railway and the Federal Export Tax. Economic Record (September), 235-41.

[31] Che, Y.-K. and S.-W. Yoo (2001). Optimal incentives for teams. American Economic Review 91(3), 525-541.

[32] Chen, K.-P. (2003). Sabotage in promotion tournaments. Journal of Law, Economics and Organization 19, 119-140.

[33] Coates, D. (1993). Property tax competition in a repeated game, Regional Science and Urban Economics 23, 111-119.

[34] Commission of the European Communities (1992). Report of the committee of independent experts on company taxation.

[35] de Haan, J. and J.-E. Sturm (1997). Political and economic determinants of OECD budget deficits and government expenditures: A reinvestigation. European Journal of Political Economy 13, 739-750.

[36] Devereux, M. P., R. Griffith and A. Klemm (2002). Corporate income tax reforms and international tax competition. Economic Policy 17, Issue 35, 449-495.

[37] Devereux, M. P., B. Lockwood and M. Redoano (2008). Do countries compete over corporate tax rates? Journal of Punlic Economics 92(5-6), 1210-1235.

[38] Dewatripont, M., I. Jewitt and J. Tirole (1999a). The economics of career concerns, part I: Comparing information structures. Review of Economic Studies 66, 183-198.

[39] Dewatripont, M., I. Jewitt and J. Tirole (1999b). The economics of career concerns, part II: application to missions and accountability of government agencies. Review of Economic Studies 66, 199-217.

[40] Dewatripont, M. and P. Seabright (2006). "Wasteful" public spending and state aid control. Journal of the European Economic Association 4(2-3), 513-522.

[41] Downs, A. (1957). An economic theory of democracy. New York: Harper & Row.

[42] Drazen, A. (2000). Political Economy in Macroeconomics. Princeton, NJ: Princeton University Press.

[43] Duverger, M. (1954). Political parties. London: Methuen.

[44] Edwards, J. and M. Keen (1996). Tax competition and Leviathan. European Economic Review 40, 113-134.

[45] Fearon, J. D. (1999). Electoral accountability and the control of politicians: Selecting good types versus sanctioning poor performance. In Przeworski, Stokes & Manin (Eds.), Democracy, accountability and representation. Cambridge University Press.

[46] Ferejohn, J. (1986). Incumbent performance and electoral control. Public Choice 50, 5-25.

[47] Flowers, M. R. (1988). Shared tax sources in a Leviathan model of federalism. Public Finance Quarterly 16, 67-77.

[48] Friedman, J. W. (1971). A non-cooperative equilibrium for supergames. Review of Economic Studies 38, 1-12.

[49] Fuest, C., B. Huber and J. Mintz (2005). Capital mobility and tax competition: A survey. Foundations and Trends in Microeconomics 1(1), 1-62.

[50] Geys, B. (2006). Looking across Borders: A Test of Spatial Policy Interdependence using Local Government Efficiency Ratings. Journal of Urban Economics 60(3), 443-462.

[51] Geys, B., B. Heyndels and J. Vermeir (2006). Explaining the formation of minimal coalitions: Anti-system parties and anti-pact rules. European Journal of Political Research 45, 957-984.

[52] Haufler, A. (2001). Taxation in a global economy: Theory and evidence. Cambridge University Press.

[53] Haufler, A. and G. Schjelderup (2000). Corporate tax systems and cross country profit shifting. Oxford Economic Papers 52, 306-325.

[54] Heyndels, B. and J. Vuchelen (1998). Tax mimicking among Belgian municipalities. National Tax Journal 51, 89-101.

[55] Hillman, A. and J. Riley (1989). Politically contestable rents and transfers. Economics and Politics 1, 17-40.

[56] Hirschleifer, J. and J. Riley (1992). The Analytics of Uncertainty and Information. Cambridge University Press.

[57] Holmström, B. (1982a). Moral hazard in teams. Bell Journal of Economics 13(2), 324-340.

[58] Holmström, B. (1982b). Managerial incentive problems - A dynamic perspective. In: Essays in Economics and Management in Honor of Lars Wahlbeck. Helsinki: Swedish School of Economics. Reprinted in Review of Economic Studies 66 (1999), 169-182.

[59] Hong, C. S. and K. A. Konrad (1998). Bandwagon effects and two-party majority voting. Journal of Risk and Uncertainty 16(2), 165-172.

[60] Hotelling, H. (1929). Stability in Competition. Economic Journal 39, 41-57.

[61] Illera, R. M. and C. Mulas-Granados (2008). What makes fiscal consolidations last? A survival analysis of budget cuts in Europe (1960-2004). Public Choice 134, 147-161.

[62] Itoh, H. (1991). Incentives to help in multi-agent situations. Econometrica 59, 611-636.

[63] Kanbur, R. and M. Keen (1993). Jeux sans frontières: tax competition and tax coordination when countries differ in size, American Economic Review 83(4), 877-892.

[64] Keen, M. (1998). Vertical tax externalities in the theory of fiscal federalism. IMF Staff Papers 45(3), 454-485.

[65] Keen, M. and C. Kotsogiannis (2002). Does federalism lead to excessively high taxes? American Economic Review 92(1), 363-370.

[66] Keen, M. and M. Marchand (1997). Fiscal competition and the pattern of public spending. Journal of Public Economics 66, 33-53.

[67] Kessing, S. G., K. A. Konrad and C. Kotsogiannis (2006). Federal tax autonomy and the limits of cooperation. Journal of Urban Economics 59, 317-329.

[68] Key, V. O. (1966). The responsible electorate: Rationality in presidential voting 1936-60. Cambridge, MA: Harvard University Press.

[69] Kind, H. J., K. H. Midelfart-Knarvik and G. Schjelderup (2000). Competing for capital in a 'lumpy' world. Journal of Public Economics 78(3), 253-274.

[70] Konrad, K. A. (2000). Sabotage in rent-seeking contests. Journal of Law, Economics and Organization 16(1), 155-165.

[71] Konrad, K. A. (2002). Investment in the absence of property rights; The role of incumbency advantages. European Economic Review 46, 1521-1537.

[72] Konrad, K. A. (2007). Non-binding minimum taxes may foster tax competition, Manuscript, Social Science Research Center Berlin (WZB).

[73] Konrad, K. A. and G. Schjelderup (1999). Fortress building in global tax competition. Journal of Urban Economics 46, 156-167.

[74] Ladd, H. F. (1992). Mimicking of local tax burdens among neighboring counties. Public Finance Review 20, 450-467.

[75] Laver, M. and K. A. Shepsle (1990). Coalitions and cabinet government. American Political Science Review 84(3), 873-889.

[76] Laver, M. and N. Schofield (1990). Multiparty government : The politics of coalition in Europe. Oxford University Press.

[77] Lazear, E. (1989). Pay equality and industrial politics. Journal of Political Economy 97(3), 561-580.

[78] Lewis-Beck, M. S. and M. Paldam (2000). Economic voting: an introduction. Electoral Studies 19, 113-121.

[79] Ludema, R. D. and I. Wooton (2000). Economic geography and the fiscal effects of regional integration. Journal of International Economics 52, 331-357.

[80] Maskin, E. and J. Tirole (2004). The politician and the judge: accountability in government. American Economic Review 94(4), 1034-1054.

[81] McCallum, B. (1978). The political business cycle: An empirical test. Southern Economic Journal 42, 504-515.

[82] Mierau, J. O., R. Jong-A-Pin and J. de Haan (2007). Do political variables affect fiscal policy adjustment decisions? New empirical evidence. Public Choice 133, 297-319.

[83] Morris, S. (2001). Political correctness. Journal of Political Economy 109(2), 231-265.

[84] Musgrave, R. (1959). The theory of public finance. New York: McGraw-Hill.

[85] Münster, J. (2007). Selection tournaments, sabotage, and participation. Journal of Economics and Management Strategy 16(4), 943-970.

[86] Myerson, R. B. (2006). Federalism and incentives for success of democracy. Quarterly Journal of Political Science 1(1), 3-23.

[87] Nannestad, P. and M. Paldam (1994). The VP function: A survey of the literature on vote and popularity functions after 25 years. Public Choice 79, 213-245.

[88] Nordhaus, W. (1975). The political business cycle. Review of Economic Studies 42, 169-190.

[89] Norpoth, H., M. S. Lewis-Beck and J.-D. Lafay (Eds.) (1991). Economics and politics: The calculus of support. Ann Arbor: The University of Michigan Press.

[90] Oates, W. E. (1972). Fiscal Federalism. New York: Harcourt Brace Jovanovich.

[91] Persson, T., G. Roland and G. Tabellini (1997). Separation of powers and political accountability. Quarterly Journal of Economics 112 (4), 1163-1202.

[92] Persson, T. and G. Tabellini (2000). Political Economics: Explaining Economic Policy. Cambridge: MIT Press.

[93] Pigou, A. C. (1912). Wealth and Welfare. London: Macmillan.

[94] Powell, G. B. and G. D. Whitten (1993). A cross-national analysis of economic voting: taking account of the political context. American Journal of Political Science 37(2), 391-414.

[95] Rauscher, M. (1998). Leviathan and competition among jurisdictions: The case of benefit taxation. Journal of Urban Economics 44, 59-67.

[96] Revelli, F. (2005). On Spatial Public Finance Empirics. International Tax and Public Finance 12(4), 475-492.

[97] Riker, W. (1962). The theory of political coalitions. New Heaven, CT: Yale University Press.

[98] Rogoff, K. (1990). Equilibrium political budget cycles. American Economic Review 80, 21-36.

[99] Rogoff, K. and A. Siebert (1988). Elections and macroeconomic policy cycles. Review of Economic Studies 55, 1-16.

[100] Roubini, N. and J. Sachs (1989). Political and economic determinants of budget deficits in the industrial democracies. European Economic Review 33, 903-938.

[101] Samuelson, L. and G. J. Mailath (2006). Repeated Games and Reputations. Oxford University Press.

[102] Samuelson, P. A. (1954). The pure theory of public expenditures. Review of Economics and Statistics 36(4), 387-89.

[103] Schelling, T. C. (1960). The strategy of conflict. Cambridge, Mass.: Harvard University.

[104] Seabright, P. (1996). Accountability and decentralization in government: An incomplete contracts model. European Economic Review 40, 61-89.

[105] Shapiro, C. (1989). Theories of oligopoly behavior. In: R. Schmalensee and R.D. Wittig (eds.), Handbook of Industrial Organization, Volume I., Elsevier.

[106] Shleifer, A. (1985). The theory of yardstick competition. RAND Journal of Economics 16(3), 319-327.

[107] Solé-Ollé, A. (2003). Electoral accountability and tax mimicking: the effects of electoral margins, coalition government, and ideology. European Journal of Political Economy 19(4), 685-713.

[108] Solé-Ollé, A. (2006). The effects of party competition on budget outcomes: Empirical evidence from local governments in Spain. Public Choice 126, 145-176.

[109] Taagepera, R. and M. S. Shugart (1989). Seats and Votes: The Effects and Determinants of Electoral Systems. New Haven, Conn: Yale University Press.

[110] Tiebout, C. M. (1956). A pure theory of local expenditures. Journal of Political Economy 64(5), 416-424.

[111] Tufte, E. (1978). Political control of the economy. Princeton, NJ: Princeton University Press.

[112] Volkering, B. and J. de Haan, (2001). Fragmented government effects on fiscal policy: New evidence. Public Choice 109, 221-242.

[113] von Neumann, J. and O. Morgenstern (1944). Theory of games and economic behavior. Princeton, NJ: Princeton University Press.

[114] Wellisch, D. (2000). Theory of public finance in a federal state. Cambridge University Press.

[115] Wildasin, D. E. (1988). Nash equilibria in models of fiscal competition. Journal of Public Economics 35, 229-240.

[116] Wildasin, D. E. (1991). Some rudimentary 'duopolity' theory, Regional Science and Urban Economics 21, 393-421.

[117] Wildasin, D. E. (2006). Fiscal competition. In: B. R. Weingast and D. Wittman (Eds.): Oxford handbook of political economy. Oxford University Press.

[118] Wilson, J. D. (1986). A theory of interregional tax competition, Journal of Urban Economics 19, 296-315.

[119] Wilson, J. D. (1991). Tax competition with interregional differences in factor endowments. Regional Science and Urban Economics 21, 423-451.

[120] Wilson, J. D. (1999). Theories of tax competition. National Tax Journal 52, 269-304.

[121] Wilson, J. D. and D. E. Wildasin (2004). Capital tax competition: bane or boon. Journal of Public Economics 88(6), 1065-1091.

[122] Wrede, M. (1996). Vertical and horizontal tax competition: Will uncoordinated Leviathans end up on the wrong side of the Laffer curve? Finanzarchiv 53(3-4), 461-79.

[123] Zodrow, G. R. and P. Mieszkowski (1986). Pigou, Tiebout, property taxation, and the underprovision of local public goods. Journal of Urban Economics 19, 356-370.

FINANZWISSENSCHAFTLICHE SCHRIFTEN

Band 1 Werner Steden: Finanzpolitik und Einkommensverteilung. Ein Wachstums- und Konjunkturmodell der Bundesrepublik Deutschland. 1979.

Band 2 Rainer Hagemann: Kommunale Finanzplanung im föderativen Staat. 1976.

Band 3 Klaus Scherer: Maßstäbe zur Beurteilung von konjunkturellen Wirkungen des öffentlichen Haushalts. 1977.

Band 4 Brita Steinbach: "Formula Flexibility" - Kritische Analyse und Vergleich mit diskretionärer Konjunkturpolitik. 1977.

Band 5 Hans-Georg Petersen: Personelle Einkommensbesteuerung und Inflation. Eine theoretisch-empirische Analyse der Lohn- und veranlagten Einkommensteuer in der Bundesrepublik Deutschland. 1977.

Band 6 Friedemann Tetsch: Raumwirkungen des Finanzsystems der Bundesrepublik Deutschland. Eine Untersuchung der Auswirkungen der Finanzreform von 1969 auf die Einnahmenposition der untergeordneten Gebietskörperschaften und ihrer regionalpolitischen Zieladäquanz. 1978.

Band 7 Wilhelm Pfähler: Normative Theorie der fiskalischen Besteuerung. Ein methodologischer und theoretischer Beitrag zur Integration der normativen Besteuerungstheorie in der Wohlfahrtstheorie. 1978.

Band 8 Wolfgang Wiegard: Optimale Schattenpreise und Produktionsprogramme für öffentliche Unternehmen. Second-Best Modelle im finanzwirtschaftlichen Staatsbereich. 1978.

Band 9 Hans P. Fischer: Die Finanzierung des Umweltschutzes im Rahmen einer rationalen Umweltpolitik. 1978.

Band 10 Rainer Paulenz: Der Einsatz finanzpolitischer Instrumente in der Forschungs- und Entwicklungspolitik. 1978.

Band 11 Hans-Joachim Hauser: Verteilungswirkungen der Staatsverschuldung. Eine kreislauftheoretische Inzidenzbetrachtung. 1979.

Band 12 Gunnar Schwarting: Kommunale Investitionen. Theoretische und empirische Untersuchungen der Bestimmungsgründe kommunaler Investitionstätigkeit in Nordrhein-Westfalen 1965-1972. 1979.

Band 13 Hans-Joachim Conrad: Stadt-Umland-Wanderung und Finanzwirtschaft der Kernstädte. Amerikanische Erfahrungen, grundsätzliche Zusammenhänge und eine Fallstudie für das Ballungsgebiet Frankfurt am Main. 1980.

Band 14 Cay Folkers: Vermögensverteilung und staatliche Aktivität. Zur Theorie distributiver Prozesse im Interventionsstaat. 1981.

Band 15 Helmut Fischer: US-amerikanische Exportförderung durch die DISC-Gesetzgebung. 1981.

Band 16 Günter Ott: Einkommensumverteilungen in der gesetzlichen Krankenversicherung. Eine quantitative Analyse. 1981.

Band 17 Johann Hermann von Oehsen: Optimale Besteuerung. *(Optimal Taxation)*. 1982.

Band 18 Richard Kössler: Sozialversicherungsprinzip und Staatszuschüsse in der gesetzlichen Rentenversicherung. 1982.

Band 19 Hinrich Steffen: Zum Handlungs- und Entscheidungsspielraum der kommunalen Investitionspolitik in der Bundesrepublik Deutschland. 1983.

Band 20 Manfred Scheuer: Wirkungen einer Auslandsverschuldung des Staates bei flexiblen Wechselkursen. 1983.

Band 21 Christian Schiller: Staatsausgaben und crowding-out-Effekte. Zur Effizienz einer Finanzpolitik keynesianischer Provenienz. 1983.

Band 22 Hannelore Weck: Schattenwirtschaft: Eine Möglichkeit zur Einschränkung der öffentlichen Verwaltung? Eine ökonomische Analyse. 1983.

Band 23 Wolfgang Schmitt: Steuern als Mittel der Einkommenspolitik. Eine Ergänzung der Stabilitätspolitik? 1984.

Band 24 Wolfgang Laux: Erhöhung staatswirtschaftlicher Effizienz durch budgetäre Selbstbeschränkung? Zur Idee einer verfassungsmäßig verankerten Ausgabengrenze. 1984.

Band 25 Brita Steinbach-van der Veen: Steuerinzidenz. Methodologische Grundlagen und empirisch-statistische Probleme von Länderstudien. 1985.

Band 26 Albert Peters: Ökonomische Kriterien für eine Aufgabenverteilung in der Marktwirtschaft. Eine deskriptive und normative Betrachtung für den Allokationsbereich. 1985.

Band 27 Achim Zeidler: Möglichkeiten zur Fortsetzung der Gemeindefinanzreform. Eine theoretische und empirische Analyse. 1985.

Band 28 Peter Bartsch: Zur Theorie der längerfristigen Wirkungen 'expansiver' Fiskalpolitik. Eine dynamische Analyse unter besonderer Berücksichtigung der staatlichen Budgetbeschränkung und ausgewählter Möglichkeiten der öffentlichen Defizitfinanzierung. 1986.

Band 29 Konrad Beiwinkel: Wehrgerechtigkeit als finanzpolitisches Verteilungsproblem. Möglichkeiten einer Kompensation von Wehrungerechtigkeit durch monetäre Transfers. 1986.

Band 30 Wolfgang Kitterer: Effizienz- und Verteilungswirkungen des Steuersystems. 1986.

Band 31 Heinz Dieter Hessler: Theorie und Politik der Personalsteuern. Eine Kritik ihrer Einkommens- und Vermögensbegriffe mit Blick auf die Leistungsfähigkeitstheorie. 1994.

Band 32 Wolfgang Scherf: Die beschäftigungspolitische und fiskalische Problematik der Arbeitgeberbeiträge zur Rentenversicherung. Eine Auseinandersetzung mit der Kritik an der lohnbezogenen Beitragsbemessung. 1987.

Band 33 Andreas Mästle: Die Steuerunion. Probleme der Harmonisierung spezifischer Gütersteuern. 1987.

Band 34 Günter Ott: Internationale Verteilungswirkungen im Finanzausgleich der Europäischen Gemeinschaften. 1987.

Band 35 Heinz Haller: Zur Frage der zweckmäßigen Gestalt gemeindlicher Steuern. Ein Diskussionsbeitrag zur Gemeindesteuerreform. 1987.

Band 36 Thomas Kuhn: Schlüsselzuweisungen und fiskalische Ungleichheit. Eine theoretische Analyse der Verteilung von Schlüsselzuweisungen an Kommunen. 1988.

Band 37 Walter Hahn: Steuerpolitische Willensbildungsprozesse in der Europäischen Gemeinschaft. Das Beispiel der Umsatzssteuer-Harmonisierung. 1988.

Band 38 Ulrike Hardt: Kommunale Finanzkraft. Die Problematik einer objektiven Bestimmung kommunaler Einnahmemöglichkeiten in der gemeindlichen Haushaltsplanung und im kommunalen Finanzausgleich. 1988.

Band 39 Jochen Michaelis: Optimale Finanzpolitik im Modell überlappender Generationen. 1989.

Band 40 Bernd Raffelhüschen: Anreizwirkungen der sozialen Alterssicherung. Eine dynamische Simulationsanalyse. 1989.

Band 41 Berend Diekmann: Die Anleihe- und Darlehenstransaktionen der Europäischen Gemeinschaften. 1990.

Band 42 Helmut Kaiser: Konsumnachfrage, Arbeitsangebot und optimale Haushaltsbesteuerung. Theoretische Ergebnisse und mikroökonometrische Simulation für die Bundesrepublik Deutschland. 1990.

Band 43 Rüdiger von Kleist: Das Gramm-Rudman-Hollings-Gesetz. Ein gescheiterter Versuch der Haushaltskonsolidierung. 1991.

Band 44 Rolf Hagedorn: Steuerhinterziehung und Finanzpolitik. Ein theoretischer Beitrag unter besonderer Berücksichtigung der Hinterziehung von Zinserträgen. 1991.

Band 45 Cornelia S. Behrens: Intertemporale Verteilungswirkungen in der gesetzlichen Krankenversicherung der Bundesrepublik Deutschland. 1991.

Band 46 Peter Saile: Ein ökonomischer Ansatz der Theorie der intermediären Finanzgewalten – Die Kirchen als Parafisci. 1992.

Band 47 Peter Gottfried: Die verdeckten Effizienzwirkungen der Umsatzsteuer. Eine empirische allgemeine Gleichgewichtsanalyse. 1992.

Band 48 Andreas Burger: Umweltorientierte Beschäftigungsprogramme. Eine Effizienzanalyse am Beispiel des "Sondervermögens Arbeit und Umwelt". 1992.

Band 49 Jeanette Malchow: Die Zuordnung verteilungspolitischer Kompetenzen in der Europäischen Gemeinschaft. Eine Untersuchung aufgrund einer Fortentwicklung der ökonomischen Theorie des Föderalismus. 1992.

Band 50 Barbara Seidel: Die Einbindung der Bundesrepublik Deutschland in die Europäischen Gemeinschaften als Problem des Finanzausgleichs. 1992.

Band 51 Ralph Wiechers: Markt und Macht im Rundfunk. Zur Stellung der öffentlich-rechtlichen Rundfunkanstalten im dualen Rundfunksystem der Bundesrepublik Deutschland. 1992.

Band 52 Klaus Eckhardt: Probleme einer Umweltpolitik mit Abgaben. 1993.

Band 53 Oliver Schwarzkopf: Die Problematik unterschiedlicher Körperschaftsteuersysteme innerhalb der EG. 1993.

Band 54 Thorsten Giersch: Bergson-Wohlfahrtsfunktion und normative Ökonomie. 1993.

Band 55 Li-Fang Chou: Selbstbeteiligung bei Arzneimitteln aus ordnungspolitischer Sicht. Das Beispiel der Bundesrepublik Deutschland. 1993.

Band 56 Harald Schlee: Einkommensteuerliche Behandlung von Transferzahlungen. Zur Neuordnung der Familienbesteuerung sowie der Besteuerung von Versicherungsleistungen und Sozialtransfers. 1994.

Band 57 Alexander Spermann: Kommunales Krisenmanagement. Reaktionen baden-württembergischer Stadtkreise auf steigende Sozialhilfekosten und Einnahmenausfälle (1980-92). 1993.

Band 58 Otto Roloff / Sibylle Brander / Ingo Barens / Claudia Wesselbaum: Direktinvestitionen und internationale Steuerkonkurrenz. 1994.

Band 59 Claudia Wesselbaum-Neugebauer: Internationale Steuerbelastungsvergleiche. 1994.

Band 60 Stephanie Miera: Kommunales Finanzsystem und Bevölkerungsentwicklung. Eine Analyse des kommunalen Finanzsystems vor dem Hintergrund der sich abzeichnenden Bevölkerungsentwicklung am Beispiel Niedersachsens unter besonderer Berücksichtigung des Landkreises Wolfenbüttel und seiner Gemeinden. 1994.

Band 61 Wolfgang Scherf: Die Bedeutung des kaldorianischen Verteilungsmechanismus für die gesamtwirtschaftlichen Wirkungen der staatlichen Neuverschuldung. 1994.

Band 62 Rainer Volk: Vergleich der Vergünstigungseffekte der verschiedenen investitionsfördernden Maßnahmen. 1994.

Band 63 Hans-Georg Napp: Kommunale Finanzautonomie und ihre Bedeutung für eine effiziente lokale Finanzwirtschaft. 1994. 2., unveränderte Auflage 1994.

Band 64 Bernd Rahmann / Uwe Steinborn / Günter Vornholz: Empirische Analyse der Autonomie lokaler Finanzwirtschaften in der Europäischen Gemeinschaft. 1994.

Band 65 Carsten Kühl: Strategien zur Finanzierung der Altlastensanierung. 1994.

Band 66 Stephan Boll: Intergenerationale Umverteilungswirkungen der Fiskalpolitik in der Bundesrepublik Deutschland. Ein Ansatz mit Hilfe des Generational Accounting. 1994.

Band 67 Karl Justus Bernhard Neumärker: Finanzverfassung und Staatsgewalt in der Demokratie. Ein Beitrag zur konstitutionellen Finanztheorie. 1995.

Band 68 Christian Haslbeck: Zentrale versus dezentrale Internalisierung externer Effekte bei unvollständiger Information. 1995.

Band 69 Regina Müller: Horizontale oder vertikale Transfers zur Durchsetzung eines horizontalen Finanzausgleichs. 1995.

Band 70 Christian Hockenjos: Öffentliche Sportförderung in der Bundesrepublik Deutschland. Darstellung und finanztheoretische Analyse. 1995.

Band 71 Manfred Rosenstock: Die Kontrolle und Harmonisierung nationaler Beihilfen durch die Kommission der Europäischen Gemeinschaften. 1995.

Band 72 Christian Rüsch: Wohnungsbau- und Wohneigentumspolitik im Rahmen der Einkommensteuer. Eine Analyse unter steuersystematischen, verteilungspolitischen und fiskalischen Aspekten. 1996.

Band 73 Stephan Winters: Die kollektive Vorsorge für den Pflegefall im Alter. Eine Untersuchung am Beispiel der gesetzlichen Pflegeversicherung in den Niederlanden. 1996.

Band 74 Knut Blind: Allokationsineffizienzen auf Sicherheitsmärkten: Ursachen und Lösungsmöglichkeiten. Fallstudie: Informationssicherheit in Kommunikationssystemen. 1996.

Band 75 Barbara Petrick-Rump: Ökonomische Wirkungen von Steueramnestien. Untersuchung konkreter Erfahrungen ausgewählter Länder mit dem Einsatz von Steueramnestien anhand eines effizienten Steueramnestieprogramms. 1996.

Band 76 Georg Hirte: Effizienzwirkungen von Finanzausgleichsregelungen. Eine Empirische Allgemeine Gleichgewichtsanalyse für die Bundesrepublik Deutschland. 1996.

Band 77 Ulrike Kirchhoff: Die rheinland-pfälzischen Gemeinden im System des Finanzausgleichs. 1996.

Band 78 Kerstin Keil: Der soziale Mietwohnungsbau: Mängel und Alternativen. 1996.

Band 79 Bernhard Manzke: Kinderlastenausgleich versus verstärkte Einwanderung. Alternative Ansätze zur langfristigen Sicherung der Gesetzlichen Rentenversicherung. 1997.

Band 80 Hariolf M. Wenzler: Institutionenökonomik und öffentliche Finanzkontrolle. Eine Analyse am Beispiel der Europäischen Union. 1997.

Band 81 Joachim Nagel: Supply-Side Policy in den USA. Eine theoretische und empirische Analyse der angebotsorientierten Wirtschaftspolitik Reagans unter besonderer Berücksichtigung steuerlicher Aspekte. 1997.

Band 82 Heinz Lampert: Krise und Reform des Sozialstaates. 1997.

Band 83 Monika Hanswillemenke / Bernd Rahmann: Zwischen Reformen und Verantwortung für Vollbeschäftigung. Die Finanz- und Haushaltspolitik der sozial-liberalen Koalition von 1969 bis 1982. 1997.

Band 84 Berthold Fürst: Die Maastrichter Budgetkriterien im Konflikt mit der Verschuldungsautonomie der deutschen Gebietskörperschaften. 1997.

Band 85 Burkhard Pahnke: Einkommensorientierte Förderung des sozialen Mietwohnungsbaues. Bestandsaufnahme und Kritik. 1998.

Band 86 Judith Safford: Staatsverschuldung im Vereinigten Königreich. Die öffentliche Verschuldung unter der Konservativen Regierung von 1979-1994. Ursachen und Auswirkungen. 1998.

Band 87 Ralf Oberheide: Die Bekämpfung der Steuerumgehung. 1998.

Band 88 Achim Truger: Die neue Finanzwissenschaft zwischen Realitätsferne und Irrelevanz der Annahmen. Eine methodologische Analyse potentieller Verteidigungsstrategien der neuen Finanzwissenschaft gegen den Vorwurf der Realitätsferne ihres entscheidungstheoretischen Fundamentes. 1998.

Band 89 Karin Bickel: Familienbezogene Elemente im System der gesetzlichen Rentenversicherung. Unter besonderer Berücksichtigung von Ein-Eltern-Familien. 1999.

Band 90 Wolfgang Scherf: Schlüsselzuweisungen und Kreisumlage. Die Problematik der Finanzierung der Landkreise am Beispiel des kommunalen Finanzausgleichs von Rheinland-Pfalz. 1998.

Band 91 Sandra Ehmann: Familienpolitik in Frankreich und Deutschland – ein Vergleich. 1999.

Band 92 Hendrik Suermann: Einkommensteuerliche Behandlung von Währungsgewinnen und -verlusten. Eine finanzwissenschaftliche Analyse des Steuerrechts in den USA und in Deutschland. 1999.

Band 93 Rolf Bösinger: Die Neuordnung des bundesstaatlichen Finanzausgleichs 1995. Eine theoretische und empirische Analyse unter Berücksichtigung von allokationstheoretischen und polit-ökonomischen Gesichtspunkten. 1999.

Band 94 Ulrich Ermschel: Finanzwirtschaftliche Konsequenzen beim Übergang auf das Ursprungslandprinzip im Europäischen Binnenmarkt. Eine Untersuchung am Beispiel des unvollkommenen oligopolistischen Neufahrzeugmarktes. 1999.

Band 95 Ute Hansen: Überwälzte Leistungen der Administration. Eine empirische und theoretische Analyse. 2000.

Band 96 Hans-Werner Seiler: Zur Durchsetzung der Einmalbesteuerung deutscher Körperschaftsgewinne. Strategien zur Vermeidung der im deutschen Körperschaftsteuersystem angelegten Benachteiligung ausländischer Anteilseigner. Eine finanzwissenschaftliche Analyse. 2000.

Band 97 Steffen Meyer: Zwischenstaatliche Finanzzuweisungen im zusammenwachsenden Europa. Zur Gestaltung eines Finanzausgleichs für die Europäische Union. 2000.

Band 98 Marion Hübner: Ökodumping? Umweltpolitik in internationalen Oligopolmärkten. 2000.

Band 99 Christhart Bork: Steuern, Transfers und private Haushalte. Eine mikroanalytische Simulationsstudie der Aufkommens- und Verteilungswirkungen. 2000.

Band 100 Norbert Eichler: Die Probleme des Gemeindefinanzausgleichs im Kooperativen Föderalismus. Eine ökonomische Analyse am Beispiel des Bundeslandes Nordrhein-Westfalen. 2000.

Band 101 Wolfgang Scherf: Der Länderfinanzausgleich in Deutschland. Ungelöste Probleme und Ansatzpunkte einer Reform. 2000.

Band 102 Stefan Dietrich Josten: Staatsverschuldung, intertemporale Allokation und Wirtschaftswachstum. Eine theoretische Analyse staatlicher Verschuldungspolitik in Modellen exogenen und endogenen Wachstums. 2000.

Band 103 Axel Breitbach: Steuerhinterziehung und Schattenwirtschaft aus gesamtwirtschaftlicher Sicht. 2000.

Band 104 Alexander Spermann: Negative Einkommensteuer, Lohnsubventionen und Langzeitarbeitslosigkeit. 2001.

Band 105 Michael Broer: Der kommunale Finanzausgleich in Hessen. Historische Darstellung und ökonomische Analyse unter besonderer Berücksichtigung der Schlüsselzuweisungen. 2001.

Band 106 Jan-Paul Ritscher: Der Einsatz von Finanzderivaten unter einer modernisierten Schuldenstrukturpolitik des Bundes. 2002.

Band 107 Martin Gasche: Dynamische Fiskalpolitik. Makroökonomische Wirkungen der Fiskalpolitik in einem Real Business Cycle-Modell. 2003.

Band 108 Felix Brosius: Internationaler Steuerwettbewerb und Koordination der Steuersysteme. 2003.

Band 109 Claudia Hensberg: Eigennützige Regierungen im fiskalischen Wettbewerb um Kapital. 2003.

Band 110 Hans-Martin Grambeck: Konsumsteuerreformen und Konsumbesteuerung. Eine vergleichende Analyse und Bewertung verschiedener Konsumsteuermodelle unter besonderer Berücksichtigung der Probleme in offenen Volkswirtschaften. 2003.

Band 111 Antje Draheim: Probleme der finanzpolitischen Willensbildung in Europa. Eine kritische Analyse der europäischen Haushalts- und Finanzverfassung. 2004.

Band 112 Robert Nuscheler: On Competition and Regulation in Health Care Systems. 2005.

Band 113 Guido K. Raddatz: Das Eigenmittelsystem der Europäischen Union. 2005.

Band 114 Markus Euler: Ansatzpunkte für eine Reform des Finanzierungssystems der Europäischen Union. 2005.

Band 115 Marco Penske: Finanzierung der Gesetzlichen Krankenversicherung – Probleme und Reformoptionen. 2006.

Band 116 Pascal Krimmer: Demographie, Erwerbsbeteiligung, Steuerreformen und Steueraufkommen. Langfristige Auswirkungen von Steuerreformen. 2008.

Band 117 Frank Blasch: Steuerreformen und Unternehmensentscheidungen. Eine empirische Analyse der deutschen Steuerpolitik mit besonderem Schwerpunkt auf die Steuerreform 2000. 2008.

Band 118 Tobias Arens: Inter- und intragenerative Umverteilung im deutschen Steuer-Transfer-System. Langfristige Wirkungen im Lebenszyklus. 2009.

Band 119 Áron Kiss: Essays in Political Economy and International Public Finance. 2009.

www.peterlang.de